Maintenance Management

Edited by Fausto Pedro García Márquez and Mayorkinos Papaelias

Published in London, United Kingdom

IntechOpen

Supporting open minds since 2005

Maintenance Management
http://dx.doi.org/10.5772/intechopen.79248
Edited by Fausto Pedro García Márquez and Mayorkinos Papaelias

Contributors
Milton Fonseca Júnior, Tirso Lorenzo Reyes Carvajal, Jandecy Cabral Leite, David Barbosa de Alencar, Jorge De Almeida Brito de Almeida Brito Júnior, Manoel Henrique Reis Nascimento, Carlos Alberto Alberto Oliveira De Oliveira de Freitas, Jorge Laureano Moya Rodriguez, Ubiratan Holanda Holanda Berzerra, Nadime Mustafa Moraes, Antônio Estanislau Sanches, Zineb Aman, Latifa Ezzine, Haj El Moussami, Sergey Kirillov, Aleksandr Kirillov, Vitalii Yakimkin, Michael Pecht, Yan-Guo Wang, Dapeng Xie, Yi Zhang, Wei Zhou, Xiantang Xue, Wenxuan Zhang, Kai Tao, Qiang Han, Kodji Deli, Djongyang Noel, Tojiddin Khayrullaevich Juraev, Riaz Khan, Ammeran B. Mad, Khairil Osman, Mohd Asyraf Abdul Aziz, Igor Timoshenko, Leila Moradi, Reza Ehteshamrasi, Fausto Pedro García Márquez, Mayorkinos Papaelias

First published in London, United Kingdom, 2020 by IntechOpen
IntechOpen is the global imprint of INTECHOPEN LIMITED, registered in England and Wales, registration number: 11086078, 7th floor, 10 Lower Thames Street, London, EC3R 6AF, United Kingdom
Printed in Croatia

British Library Cataloguing-in-Publication Data
A catalogue record for this book is available from the British Library

Additional hard and PDF copies can be obtained from orders@intechopen.com

Maintenance Management
Edited by Fausto Pedro García Márquez and Mayorkinos Papaelias
p. cm.
Print ISBN 978-1-83968-513-2
Online ISBN 978-1-83968-514-9
eBook (PDF) ISBN 978-1-83968-515-6

For EU product safety concerns:
IN TECH d.o.o., Prolaz Marije Krucifikse Kozulić 3, 51000 Rijeka, Croatia, info@intechopen.com or visit our website at intechopen.com.

Meet the editors

Fausto Pedro García Márquez (https://blog.uclm.es/faustopedrogarcia/) works at UCLM as a full professor (accredited as full professor from 2013), Spain, an honorary senior research fellow at Birmingham University, UK, and a lecturer at the Postgraduate European Institute; he has also been a senior manager at Accenture (2013–2014). He obtained his European PhD with a maximum distinction. He has been distinguished with the following prizes: Advancement Prize for Management, Science and Engineering Management Nominated Prize (2018), First International Business Ideas Competition 2017 Award (2017); Runner-up (2015), Advancement (2013), and Silver (2012) from the International Society of Management Science and Engineering Management (ICMSEM); and Best Paper Award in the *International Journal of Renewable Energy* (impact factor 3.5) (2015). He has published more than 150 papers (65% ISI, 30% JCR, and 92% international), with some recognized as: "Renewable Energy" (Best Paper 2014); "ISMSEM" (excellent), "Int. J. of Automation and Computing" and "IMechE Part F: J. of Rail and Rapid Transit" (most downloaded), etc. He is author and editor of 25 books (Elsevier, Springer, Pearson, Mc-GrawHill, IntechOpen, IGI, Marcombo, AlfaOmega, etc.), and five patents. He is editor of five International journals and committee member of more than 40 international conferences. He has been principal investigator in four European projects, five national projects, and more than 150 projects for universities, companies, etc. His main interests are maintenance management, renewable energy, transport, advanced analytics, and data science. He is a director of the Ingenium Research Group (www.ingeniumgroup.eu).

Dr. Mayorkinos Papaelias (PhD in Metallurgy, chartered engineer, Greece) is a senior lecturer in non-destructive testing (NDT) and condition monitoring at the School of Metallurgy and Materials at the University of Birmingham. Dr. Papaelias is an expert in NDT and condition monitoring technology. He has been involved as a technical coordinator or scientific consultant in several FP6 and FP7 collaborative research projects. He is the author or coauthor of more than 70 journal and conference papers in NDT and condition monitoring. He is also a member of the International Society for Condition Monitoring. Dr. Papaelias regularly authors articles for industrial magazines.

Contents

Preface

Maintenance is a critical variable in industry to achieve competitiveness. Therefore, correct management of corrective, predictive, and preventive politics in any industry is required. *Maintenance Management* considers the main concepts, state of the art, advances, and case studies in this topic. This book complements other subdisciplines such as economics, finance, marketing, decision and risk analysis, engineering, etc.

The book analyzes real case studies in multiple disciplines. It considers the topics of failure detection and diagnosis, fault trees, and subdisciplines (e.g. FMECA, FMEA, etc.). It is essential to link these topics with finance, scheduling, resources, downtime, etc. to increase productivity, profitability, maintainability, reliability, safety, and availability, and reduce costs and downtime.

This book presents important advances in mathematics, models, computational techniques, dynamic analysis, etc., which are all employed in maintenance management.

Computational techniques, dynamic analysis, probabilistic methods, and mathematical optimization techniques are expertly blended to support the analysis of multicriteria decision-making problems with defined constraints and requirements.

The book is ideal for graduate students and professionals in industrial engineering, business administration, industrial organization, operations management, applied microeconomics, and the decisions sciences, either studying maintenance or who are required to solve large, specific, and complex maintenance management problems as part of their jobs. The book will also be of interest to researchers from academia.

The editors are very grateful to all the authors who have contributed to this book.

Fausto Pedro García Márquez
Ingenium Research Group,
Universidad Castilla-La Mancha,
Ciudad Real, Spain

Mayorkinos Papaelias
School of Metallurgy and Materials,
University of Birmingham,
Birmingham, United Kingdom

Chapter 1

Introductory Chapter: An Overview to Maintenance Management

Fausto Pedro García Márquez and Mayorkinos Papaelias

1. An overview to maintenance management

The industry requires maintenance to ensure the correct operations of the engines, components, structures, etc. [1] Any failure, that is, termination of the ability of an item to perform a required function, generates downtimes, costs, risks for the human labors, etc. The high competitiveness in the current industry does not lead these failures to the firms [2].

The advances in information and communication systems, together with the technologies, lead to the industry to incorporate new sensors, condition monitoring systems, etc. [3] They also require advance analytics in order to format, save, and analyze these signals and information, from qualitatively and quantitative point of views [4].

In order to reduce the failures occurrence probability, a correct maintenance task is required. British Standard, BS EN-13306:2017 [5] defines maintenance as "managerial actions during the life cycle of an item intended to retain it in, or restore it to, a state in which it can perform the required function. Technical maintenance actions include observation and analyses of the item state (e.g., inspection, monitoring, testing, diagnosis, prognosis, etc.) and active maintenance actions (e.g., repair, refurbishment)." The correct maintenance support to a maintenance organization to carry out the correct tasks is called maintenance supportability.

There are a large number of maintenance types, where the principal could be:

- **Corrective maintenance**, is the most common type, is done when the failure appears. In case, if it is delayed, it is defined as deferred corrective maintenance; in other case, it is called as immediate corrective maintenance.

- **Preventive maintenance**, done in certain times or according to criteria to reduce the failure probability [6]. Predetermined maintenance is set according to intervals of times or use of the item. Scheduled maintenance is done as predetermined maintenance or in a time schedule established previously. Condition-based maintenance is carried out regarding to the item status that is set generally by sensors, testing, and analytics [7].

- **Predictive maintenance**, the maintenance tasks are done according to the item condition predicted in order to avoid a failure [8].

Figure 1 shows the main maintenance types.

IntechOpen

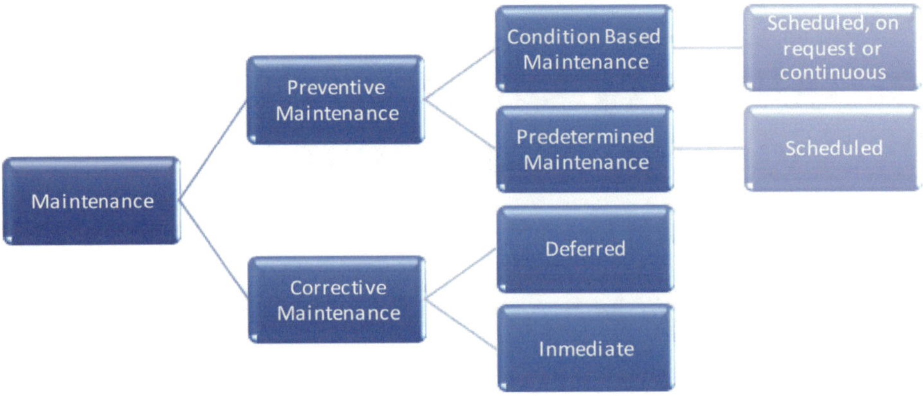

Figure 1.
Overall view of maintenance types [5].

According to EN 13306:2010 [9], maintenance management is defined as "all activities of the that determine the maintenance objectives, strategies and responsibilities, and implementation of them by such means as maintenance planning, maintenance control, and the improvement of maintenance activities and economics." The maintenance strategy is set to get the objectives, fixed by the costs, availability, safety, reliability, etc. The maintenance strategy should be set by maintenance management by the responsibility point of view, considering the availability, safety of the human, the environment, and any other mandatory requirements associated with the item, etc., item durability and the final product quality taking into account the cost, and any influence to the environment [10]. The procedures, activities, resources, and time are considered in the maintenance plan structured.

The main key indicators are found in European Standard EN 15341:2007 [11]. The objectives of the key indicators are to measure the status, compare (internal and external benchmarks), diagnose (analysis of strengths and weaknesses), identify objectives, and define targets to be reached, plan improvement actions, and continuously measure changes over time. There are three main groups of indicators: economic [12], technical [13], and organizational [14]. They are set considering endogenous (company culture, industry, life cycle of the components, criticality, etc.) and exogenous (location, society culture, market, laws, regulations, etc.) variables [15].

Author details

Fausto Pedro García Márquez[1*] and Mayorkinos Papaelias[2]

1 Ingenium Research Group, Universidad Castilla-La Mancha, Ciudad Real, Spain

2 School of Metallurgy and Materials, University of Birmingham, United Kingdom

*Address all correspondence to: faustopedro.garcia@uclm.es

References

[1] Muñoz CQG, Márquez FPG. Future maintenance management in renewable energies. In: Renewable Energies. Springer; 2018. pp. 149-159

[2] Marugan AP, Marquez FPG. Decision-Making Management: A Tutorial and Applications. Academic Press; 2017

[3] Pérez JMP, Márquez FPG, Papaelias M. Techno-economical advances for maintenance management of concentrated solar power plants. In: Proceedings of the Tenth International Conference on Management Science and Engineering Management. Springer; 2017. pp. 967-979

[4] Márquez FPG, Lev B. Advanced Business Analytics. Springer; 2015

[5] BS_EN_13306:2017. Maintentance–Management Terminology. BSI Standards Publication; 2017

[6] Pliego A, de la Hermosa RR, Márquez FPG. Big data and wind turbines maintenance management. In: Renewable Energies. Springer; 2018. pp. 111-125

[7] Márquez FPG, Segovia I. Condition monitoring system for solar power plants with radiometric and thermographic sensors embedded in unmanned aerial vehicles. Measurement. 2019

[8] Gómez Muñoz CQ, García Marquez FP, Hernandez Crespo B, Makaya K. Structural health monitoring for delamination detection and location in wind turbine blades employing guided waves. Wind Energy. 2019

[9] EN_13306:2010. Maintenance–Maintenance Terminology. European Committee for Standardization; 2010

[10] Marugán AP, Márquez FPG. Decision making approach for optimal business investments. In: Advanced Business Analytics. Springer; 2015. pp. 1-20

[11] EN_15341:2007. Maintenance–Maintenance Key Performance Indicators. European Standard; 2010

[12] Pérez JMP, Asensio ES, Márquez FPG. Economic viability analytics for wind energy maintenance management. In: Advanced Business Analytics. Springer; 2015. pp. 39-54

[13] Jiménez AA, Muñoz CQG, Márquez FPG. Machine learning and neural network for maintenance management. In: International Conference on Management Science and Engineering Management. Springer; 2017. pp. 1377-1388

[14] Marugán AP, Márquez FPG. Improving the efficiency on decision making process via bdd. In: Proceedings of the Ninth International Conference on Management Science and Engineering Management. Springer; 2015. pp. 1395-1405

[15] Pliego Marugán A, García Márquez FP, Lev B. Optimal decision-making via binary decision diagrams for investments under a risky environment. International Journal of Production Research. 2017;**55**:5271-5286

Chapter 2

Remote Computing Cluster for the Optimization of Preventive Maintenance Strategies: Models and Algorithms

Aleksandr Kirillov, Sergey Kirillov, Vitaliy Iakimkin and Michael Pecht

Abstract

The chapter describes a mathematical model of the early prognosis of the state of high-complexity mechanisms. Based on the model, systems of recognizing automata are constructed, which are a set of interacting modified Turing machines. The purposes of the recognizing automata system are to calculate the predictors of the sensor signals (such as vibration sensors) and predict the evolution of hidden predictors of dysfunction in the work of the mechanism, leading in the future to the development of faults of mechanism. Hidden predictors are determined from the analysis of the internal states of the recognizing automata obtained from wavelet decompositions of time series of sensor signals. The results obtained are the basis for optimizing the maintenance strategies. Such strategies are chosen from the classes of solutions to management problems. Models and algorithms for self-maintenance and self-recovery systems are discussed.

Keywords: turing machine, maintenance optimization, preventive maintenance, remaining useful life, remote calculating cluster

1. Introduction

This chapter describes a mathematical model that allows to unify the multiplicity of approaches to the creation of intelligent maintenance systems on the one hand and also allows more to accurately formalize and then algorithmize the optimization tasks of maintenance strategies.

Consideration of problems in the management of maintenance is useful to begin with the formalization of the basic tasks of PHM. The main trend of today is the development of prognostics and health management (PHM) in the sequence condition-based maintenance (CBM)-predictive maintenance (PdM) [1]. This concept may be called "CBM +" or "proactive management of materials degradation," and then on the horizon, there are new concepts, bearing a semantic load, in particular self-maintenance and self-recovery systems. However, all these concepts need further formalization (mathematical). It should also be noted that the effectiveness of any maintenance strategy depends on how reliably the PHM system is able to predict the state of technical objects of high complexity. Construction of the effective maintenance strategies is possible on the basis of a reliable prognosis.

Let us dwell in more detail on this question. Any prognostic system is based on the statistical processing of the signals received by sensors mounted on the technical objects. This can be a variety of vibration sensors, sensors of pressure, and measurement of currents and voltages. In this chapter, we will appeal to the examples of prognosis of the technical state of rotating machinery and the reciprocating action mechanisms, demonstrating the commonality of models.

Also, the chapter will pay attention to medical applications, in particular, to remote cardiac monitoring systems. Here, the task of the prognosis consists of three sub-tasks: a prognosis of the state of the heart on the basis of wearable or implanted in the body-miniaturized ECG recorders; construction of the management model of the heart state with the help of variable parameters of implantable devices in the human body, such as ICD and CRT devices; and prognosis and estimation of the remaining useful life (RUL) of the implanted devices [2, 3]. It should be noted that not only statistical methods are the basis of the prognosis, but also, more importantly, in this basis, physical models of the monitored object and its subsystems should be contained. Ultimately, we are talking about digital counterparts, that is, accounting for all components and processes occurring in a working device is necessary.

Thus, there is a task of prognosis of the technical state of the object and a time estimation under the general name remaining useful life (RUL). The prognosis of the technical condition and RUL estimation are the bases for constructing cost-effective maintenance strategies.

It is on the basis of the prognosis and RUL estimates is possible formalization of the task of determining the cost-effective maintenance strategies, taking into account the conditions of goal setting. Goal setting involves taking into account the requirements for the technology and determining the ultimate goal of its operation. For example, in monitoring vehicles, extending the time of operation with a minimum change in operating parameters, for example, the engine and its subsystems, is a natural requirement that determines the strategy for calculating optimal operating conditions. Obviously, the calculated strategy is unsatisfactory for military applications, since it will explicitly prohibit operation with violation of optimal speed regimes and all kinds of extreme exploitation. When calculating maintenance strategies for military systems, the goal setting conditions change, where the determining strategy is the delivery of one's own weapons to a given point of space at any cost, taking into account the impact of enemy-striking factors.

The noted condition directly points to the fact that a maintenance strategy with necessity must be determined in a number of cases in real-time conditions, while the goal setting itself will change during operation. The transition to earlier prediction methods that can be called the diagnosis or prognosis of the root causes, hidden predictors of prognosis, etc. creates the conditions to search for more effective maintenance strategies. And, finally, the creation of self-maintenance and self-recovery systems requires the presence of a physical model of processes, within which functional dependencies between the parameters of management of process and its state are determined.

2. Basis

For precise algorithmized formulations of optimization tasks for maintenance strategies, mathematically rigorous formalizations of the basic concepts of prediction tasks and tasks of management of the state of technical object are necessary.

We will assume that the system is equipped with all the necessary sensors, registering the vibration of the engine housing, the sensors of the angle of rotation of the shaft or crankshaft, pressure sensors in high-pressure fuel lines and other necessary sensors, most of which are included in the system of traditional onboard diagnostics or control systems. It also means the possibility of transmitting sensor signals (time series) to a remote computing cluster.

Further input to the computing cluster signals or time series is presented in the form of their wavelet coefficients. The fixation of all indices of the wavelet coefficients except for the quasi-period index, that is, the current number of the cycle of turbine engine, etc., determines the so-called cascades. The entire set of cascades is considered. Their number is equal to the number of wavelet coefficients of the decomposition of the time series multiplied by the number of sensors from which signals are received. A set of finite segments of fixed cascade defines a state vector in its sequence. The evolution of the state vector at successive change of segments determines the vector of trajectory. The multi-trajectory is determined by the vector of trajectories of all cascades, that is, a set of state vectors determines the multistate or state of the entire system.

The first prognosis problem is reduced to the definition of the evolution equations describing the evolution of the state vector.

Depending on the properties of the process, these equations are known in the sections of nonequilibrium thermodynamics called the "basic kinetic equation." The basic kinetic equation is reduced depending on the properties of the cascade (stationarity, ergodicity, nonstationarity, Markovity, non-Markovity, etc.) and reduces to equations such as the Fokker-Planck equation, the Schrodinger equation, the balance equation, to single-step processes, etc. [4, 5].

The prognosis task is formulated as a definition of the probability of a transition from the initial state vector to the final one and preassigned [6, 7], for example, preassigned on the boundary of the failure region, on the boundary of the region of the nucleation of failure predictor, or on the boundary of failure predictors. The development or evolution of predictors or hidden predictors is also described by evolutionary equations of the type listed above. Thus, there is a set of trajectories or multi-trajectories of sequences of states of the system. Further formalization requires the classification of trajectories in order to determine the trajectories leading to the boundaries of failure. The boundaries between classes of different trajectories may not be physically observable. However, these boundaries affect the trajectories, changing their characteristics. For example, in the case of interpreting a trajectory as a random walk in a multidimensional lattice or its continual counterpart, the evolution equations themselves and, consequently, the RUL estimates change.

Thus, there arises the problem of classifying a set of physically feasible trajectories or the task of representing trajectories in the form of a set of classes and the task of describing the boundaries between classes. Separation of the set of trajectories into classes is a rather ambiguous task, and often there are problems with changing the classification when changing the types of processes. However, it follows from the constructed model that the separation of trajectories into classes is related to the transformation of the topological characteristics of the state space and trajectory spaces.

In the case under consideration, each class is characterized by its own group of symmetries of the probability density of transitions between vectors and/or the group of symmetries of the generating functional. Factorization of the symmetry group by the isotropy subgroup, leaving the vector state in place, generates a homogeneous space [8]. It is in this space that the vector process wanders. In the process of operation of the mechanism and degradation of the material, the topology of the homogeneous space is changed; this change generates the boundary between classes.

Let us return to the tasks of management. In the concepts defined above, the task of management is formalized as follows in which variations of the controlled parameters preserve the trajectory of the states of the system in the given class for as long as possible. The following formulation concerns the estimates of RUL as an estimate of the time to reach the class boundary when the controlled parameters are varied. When crossing class boundaries, the task of evaluating the RUL and maximizing the time of stay in the class is solved again. In this case, the evolution equations change.

The model described above makes it possible to formalize the problem of finding optimal maintenance strategies as a task of determining control parameters or more precisely determining the range of admissible control parameters under which the trajectory is kept as long as possible in a given class.

3. Model

The search of a way to formalize maintenance management tasks and building models and algorithms for searching for optimizing strategies is useful to start with the formalization of the management process in an extremely general setting. The approach presented here is rather complicated, but it is useful for the development of further formalizations and construction of algorithms.

To do this, we will present the task of managing, using the following definitions. Let the considered technical object have in its arsenal several parameters, the variation of which affects the state of the mechanism, changing all the permissible modes of its operation. More precisely, the variation of the control parameters allows the mechanism to be switched from one operating mode to another physically acceptable mode. Next, consider some abstract mathematical space; often, these are certain subsets of a multidimensional space $\boldsymbol{R}^{N^*}$. Further constructions show that these subsets of space $\boldsymbol{R}^{N^*}$ are topological manifolds with a complex topology.

Each point of such space determines the state of the mechanism at a fixed time; the sequence of states defines a trajectory in the state space. We also accept, as an empirically understandable assumption, that when the control parameters are varied, the continuous trajectories change in the same way without discontinuities. That is, a small perturbation of the parameters also causes a slight perturbation of the trajectory; in other words, for small perturbations the new trajectory is in some sense close to the original trajectory.

As a result, a continuous mapping from the parameter space to the state space is determined. **Figure 1** demonstrates the mapping of the management loop Ω, consisting of two management parameters to the state space. It is assumed that all values of the parameters inside the circuit are physically realizable. When mapping the management interval $I = \{\lambda_i\} \stackrel{\text{def}}{=} [0, 1]$, the path is formed from the initial to the final state in the state space. As a result, at the change of parameters and with changes in the state of the mechanism during operation, many paths are generated. The set of paths in the state space **X** defines a new space [9], designated as $\Omega\boldsymbol{X}$—the loop space of space $\boldsymbol{X}$. In this case, the next parameter determines already the mapping of the management interval:

$$I \to \boldsymbol{\Omega X} \tag{1}$$

thus defining a twofold loop space $\Omega^2\boldsymbol{X}$ in the space $\boldsymbol{X}$.

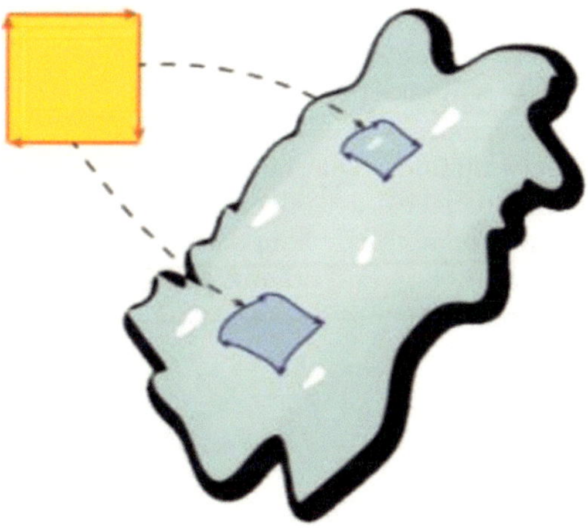

Figure 1.
Mapping the management loop $X\ \mathbf{R}^n \times \mathbf{R}^1$ *in state space;* $\lambda_1 \ldots \lambda_n$;*—Variable parameters,* t-time.

$$\Omega^2 \boldsymbol{X} \stackrel{\text{def}}{=} \Omega(\Omega \boldsymbol{X}) \tag{2}$$

$$\Omega^k \boldsymbol{X} \stackrel{\text{def}}{=} \Omega^{k-1}(\Omega \boldsymbol{X}) \tag{3}$$

Further presentation will require some information from algebraic topology, more precisely the homotopy theory. In view of the complexity of mathematical constructions, one must sacrifice a mathematically rigorous exposition in favor of simplification and clarity. Thus, the above arguments necessarily lead to an analysis of the set of paths $\Omega \boldsymbol{X}$ in the state space $\boldsymbol{X}$ defined by the mapping Eq. (1) for one management parameter. With an increasing number of parameters, the path space is generated in the path space of the previous parameter Eq. (3), and so on with the growth of the number of control parameters. As a result, taking into account all control parameters leads to consideration of the k-fold space of paths, more precisely to the k-fold loop space [9].

Returning to the basic concepts of homotopy theory, it should be noted that the methods mentioned here were used in the 1970s in the physics of a condensed state for the analysis of singularities in condensed media, including superconductors (Abrikosov vortices), superfluid liquids, and liquid crystals. The methods of homotopy topology are effective not only for general analysis and classification of singularities of condensed media [10] but also transferred to the analysis of processes expressed in the form of multiple spaces of loops. This fact can be explained as follows. The management contour at the mapping to the state space defines a contour in the state space itself or on the corresponding loop space, the multiplicity of which is determined by the number of management parameters. The following problem arises, solved by the homotopy theory methods. Can the image ∂I^k under the mapping and defined by the contour in the state space or constructed Eq. (4, 5) on it the loop space be continued from the boundary of the set I^k, ∂I^k, to its interior I^k in a continuous manner? Or such continuation is impossible, that means the presence of topological obstacles, expressed by the nontriviality of the topological (homotopy) type of the state space, the loop space. In the case of obstacles, any continuation will undergo a discontinuity in the corresponding topology of the loop space. In the case when the mapping F to the loop space is topologically nontrivial, that is, corresponds to a nontrivial element of the homotopy group of the state space or loop spaces, then a discontinuity will occur when the management parameters are varied. This means that it is not possible to continue the regularity from the boundary of the management loop to its interior without discontinuities. Physically, with small variations in the management parameters, a transition from the initial process to the final process will take place abruptly. This, depending on the specific physical content of the management model, leads to dramatic changes in the state of the mechanism that is accompanied by a sharp change in the operating conditions and extreme loads, leading to accelerated degradation of the material: the nucleation and growth of microcracks, the development of abnormal wear in the corresponding mechanical junction and other troubles, the precursor of creation of avalanche changes in the material, and so on.

$$F : \partial I^k \to X \tag{4}$$

$$I^k = \{\lambda_i : i = 1, ...k\} \stackrel{\text{def}}{=} \prod_k [0,1] \tag{5}$$

The above results need more precise definitions of the state space X, the space of trajectories, and the identification of physical causes for the appearance of homotopically nontrivial state space.It is appropriate here again to use the analogy with topological defects of condensed media. It has already been noted above that when using analogies of this kind, it is only necessary to redefine the notion of a degeneracy space. The redefined degeneracy space in this case and thanks to the work [9] is nothing more than a k-fold loop space.

Topological singularities in condensed media are provided by the homotopy nontriviality of the so-called degeneracy space of the free energy functional of a condensed medium. The presence of the degeneracy group of the free energy and its further factorization with respect to the isotropy subgroup gives the required degeneration space, in mathematics called the homogeneous space [10]. In the task under consideration, the analog of the construction of the degeneracy space is in the most general case the characteristic functional of the stochastic process. The symmetry groups of such a functional are considered in [11]. To understand the methods of constructing degeneration spaces, one can consider the density of function of the distribution of the process. If we return to the cascades of the wavelet coefficients of the observed signal and then to the vector processes, then we consider the vector process or segments of length N or the set of such segments or vectors under certain assumptions about the properties of the observed process, for example, if the process reduces to a random walk in a multidimensional lattice or on a continuum. In the example under consideration, the group of probability density function (PDF) of the process has a Gaussian distribution, and hence the symmetry group of such a process is the group $SO(N)$.

For example, in the problem of walk of $\boldsymbol{R}^N$ [6, 12], the Gaussian function for the density of probability of falling into a point $R \in \boldsymbol{R}^N$ after traversing the path of length L is the following:

$$G(R;L) = \left(\frac{N}{2\pi lL}\right)^{\frac{N}{2}} \exp\left(-\frac{N\|R\|^2}{2lL}\right). \tag{6}$$

The subgroup of isotropy is in this case the subgroup of rotations of the vector R about its axis, that is, $SO(N-1)$. The result of the factorization $SO(N)$ of the group with respect to the subgroup $SO(N-1)$ is the N-1-dimensional sphere:

$$SO(N)/_{SO(N-1)} \cong S^{N-1} \tag{7}$$

Taking into account other symmetries existing in the observed process changes the degeneracy space. For example, the vector processes under consideration can in some cases have symmetry with respect to time reversal, and then the vector field becomes a field of directors as in a nematic liquid crystal. In this case, the degeneracy space is transformed from a sphere into a projective space of dimension:

$$SO(N)/_{SO(N-1)\times Z_2} \cong PR^{N-1} \tag{8}$$

The presence of trends or dynamic predictors removes such degeneracy, and the degeneracy space again becomes a sphere. The permutation group acting on the components of the vectors, that is, changes their places, turns the sphere into an even more complex homogeneous space, where the gluing takes place in the discrete orbits of the group of permutations during factorization, generating a space homotopically equivalent to a bouquet of spheres of different dimensions (**Figure 2**):

$$\mathrm{DS} = \mathrm{S}^{\mathrm{N}-1} \bigvee\nolimits_{\{\mathrm{i}\}} \mathrm{S}^1 \tag{9}$$

The transition to space trajectories (spaces of k-fold loops) determines in the final analysis ultimately a classification of trajectories, representing each class from the set of admissible trajectories as a set of homotopy equivalent trajectories. The set of homotopy classes of such spaces is denoted as in [9]. This set has a structure of group as follows from the given examples.

Useful relations for computing homotopy groups of homogeneous spaces and loop spaces are given below, along with examples of homotopy groups of spheres and other homogeneous spaces:

$$[W, \Omega X] \leftrightarrow [\Sigma W, X] \tag{10}$$

$$\pi_i(\Omega X) \cong \pi_{i+1}(X) \tag{11}$$

$$\Sigma W \stackrel{\text{def}}{=} ((W \times I))/((W \times 0) \cup (w\_0 \times I) \cup (W \times 1) \tag{12}$$

ΣW—cited superstructure over W;

$$\pi_{n+15}(S^n) = Z_{480} \oplus Z_2 \tag{13}$$

$$\pi_1(SO^n) = Z_2 \tag{14}$$

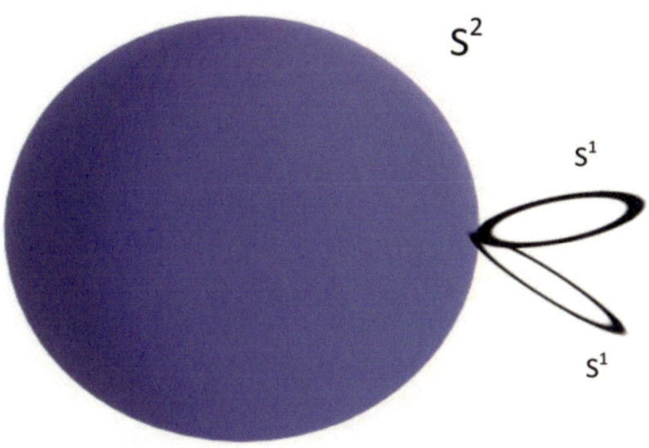

Figure 2.
A bouquet of two-dimensional and two one-dimensional spheres.

Part of the Hopf fibration is a mapping $S^3 \to S^2$ [13–15]. This mapping is the generator of the homotopy group $\pi_3(S^2) = Z$.

This is under the assumption that the degeneracy space does not change. When the homotopy type of the degeneration space changes, the classification of trajectories also changes. The change in the topology of the degeneracy space is due to a change in the symmetry groups of the process. There is a violation of symmetry due to a change in the characteristic features of the process, as already noted, for example, with the appearance of trends. Predictor of the trend is, in fact, a change in the structure in the set of transition probabilities, as will be discussed below.

Symmetry breaking or removal of degeneracy by isotropy subgroups can occur for various reasons. One such mechanism is associated with noise-induced transitions. In [16] examples of this kind are given. The reason for removing the degeneracy and, consequently, changing the topological type of the degeneracy space is the presence of multiplicative noise. As a result of the growth of the amplitude of such noise, a change occurs in the characteristics of the process, in particular, the density of the distribution function changes.

Returning to the tasks of management, the following should be noted. Thus, a space of degeneracy for the system and constructed on it k-fold loop space that is homotopy equivalent to the space of paths on the degeneracy space are defined sufficient roughly. The classification of paths is determined by the set of classes of homotopy equivalent paths. The transition from one class of paths to another class of paths is accompanied by symmetry breaking. Very conditionally the process of development of failure and dysfunctions of the mechanism can be shown in **Figure 3**. The colored concentric rings represent different types of homogeneous spaces on which it is necessary to keep the trajectory as long as possible. At the same time, the time to reach the boundary of the RUL class is estimated.

That is, the process of the development of faults as a result of operation passes from one class to another, reaching at the end of the failure field. In this case, the intersection of the conditional boundary is determined by a violation of the symmetry of the process. Further, the degeneration space itself and the character of the transition from one class to another change.

The mathematical model described above allows us to make the first step in the formulation and formalization of optimization of the maintenance strategy. The optimization task is reduced to determining the number of management parameters and determining the image of the management loop in the state space or the k-fold space of paths that hold the trajectory of the process in a given homotopy class or in a given degeneracy space. If necessary, homotopic obstacles are overcome by

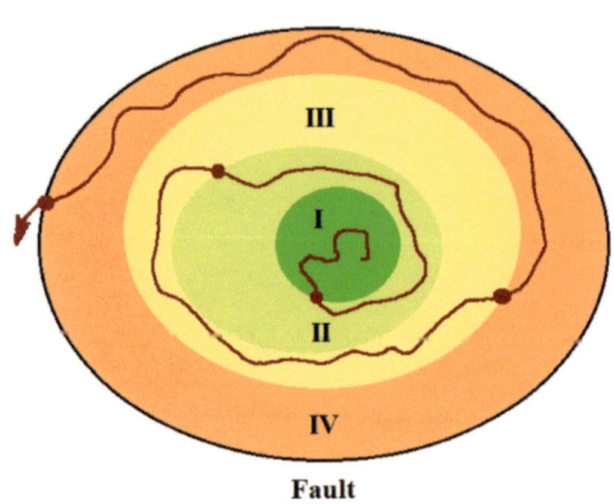

Figure 3.
Interpretation of optimization problem and maintenance strategies.

increasing the dimension of the management loop, that is, by increasing the number of management parameters. The operation of the restructuring of the degeneracy space is inevitable, therefore with each new restructuring task literally reformulated.

Restructuring degeneracy space is associated with symmetry breaking. To explain all of the above, we can use a very simplified example. Consider the degeneration group $SO(3)$ and the isotropy group $SO(2) * Z_2$. Moreover, the degeneracy space is a projective space RP_2. Removing degeneracy by Z_2 is associated with the violation of time-reversal symmetry. Such loss of symmetry is possible with the appearance of the trends of the initial cascade of wavelet coefficients of the observed time series. In this case, the degeneracy space is transformed into a sphere. Homotopy groups of projective space and spheres differ and are represented by Eqs. (15) and (16):

$$\pi_1\left(S^2\right) = 0 \tag{15}$$

$$\pi_1\left(RP^2\right) = Z_2 \tag{16}$$

The homotopy groups of k-fold loop spaces are represented by Eqs. (10) and (11). The nontriviality of homotopy groups generates a classification of paths, that is, their division into classes of homotopically nonequivalent paths in a management task with two or three management parameters.

When removing degeneracy and the transformation of the degeneration space itself into the space of another homotopy type, respectively, the homotopy classes of paths also change. In this case, new hidden predictors of failure will appear. An example is the predictors of turbine surging, described in [17]. It is noted that the early predictor of surging is destroyed by the mixing of wavelet coefficients. In this approach this means that in the observed process (the observed signal) from the pressure sensor the violation symmetry with respect to the permutation group occurred.

The above general model is based only on two statements that need concrete implementation for the further use of such model in prognosis and management tasks. Two said assumptions are as follows: there exists a space of states realized as a vector multidimensional space then taking into account the symmetry groups the set of states of the system, and the set of trajectories was represented in the form of degeneracy space and multiple path spaces that take into account the management parameters.

4. Review of the solutions of the prognosis and management task from abstract models to implementation

The constructed model on the basis of the introduced assumptions using the concepts of homotopy topology gives a general classification of admissible trajectories, their evolution. The model demonstrates the complexity of the prognosis in view of the need to take into account symmetry breaking, in other words, the removal of degeneracy by one or more subgroups of process symmetries or cascade of wavelet coefficients. That is, the description of the topological transformation of the degeneracy spaces and the association with its spaces of k-fold paths allow one to look at the tasks of prognosis and management in a different interpretation. The model allows us to describe all admissible types of topological transformations, defines classes of admissible trajectories, determines all possible transformations of classes of trajectories and subsets, and characterizes subsets in the state space and trajectory spaces related to regions of failure. Further, the same methods describe the evolution of such regions, their interaction, and pair interaction on the basis of the group structure of homotopy classes. Involving physical models from the physics of failure makes it possible to determine the physical meaning of topological nontriviality and to connect topological obstacles and topological prohibitions with physical mechanisms that ensure topological transformations.

The next step to the construction of computational algorithms for prognosis and management is the transition from topological dynamics described above to the construction of the evolution equations of trajectories and states and finally to the construction of algorithms for prognosis and management. Moreover, the conclusions and results of the homotopy model must be taken into account with necessity.

To do this, it is necessary to determine the specific content of the above concepts such as the state space and path space. A detailed exposition of this construction is contained in the works [18–23]. The observed signal is represented as the coefficients of its wavelet transformation:

$$\left\{{}^{k}_{Hist}W^{N}_{i,j}\right\}, N = 1, 2, 3, \ldots, N^{*} \tag{17}$$

N—number of cycle; i, j—indices of wavelet decomposition.

Hist—duration of cycles (unevenness of stroke) of histogram column index.

k—numbering of vectors from wavelet coefficients of dimension N^{*}.

This takes into account the fact that the mechanism under consideration is a reciprocating or rotational mechanism for all fixed indices except that N is determined by stochastic process with discrete time, N cascade. Further, fixing the limiting value N as N^{*} is determined by a set of vectors of dimension N^{*}, chosen from the consistent values of the process under consideration with discrete time. As a result, the space $\boldsymbol{R}^{N^{*}}$ is determined, consisting of all possible finite segments of dimension размерности N^{*}.

The state space is defined as follows:

$$\{\boldsymbol{R_k}\} \stackrel{\text{def}}{=} \left\{{}^{k}_{Hist}W^{N}_{i,j}: k\,N^{*} \leq N \leq (k+1)N^{*}, k = 0, 1, 2, 3\ldots\right\}, \boldsymbol{R_k} \in \boldsymbol{R}^{N^{*}} \tag{18}$$

Thus, a vector space of dimension H is defined. The numerical value of H is not yet specified. It is determined in the process of preprocessing. The task of prognosis here reduces to determining the probability of transition from an initial vector to a finite vector in j steps. In this case, such task is solvable either by an explicit solution of the evolution equations or by calculating the moments, mainly of the dispersion, that is, second moment. Similar calculations are given in [6] and are reduced in most cases to the calculation of the Feynman integral along trajectories [6, 18–24] or to the solution of evolution equations such as the Fokker-Planck equation. In this case, the trajectory of states is represented as a walk along a multidimensional lattice or its continual analog, that is, R^{N} space [6], in those cases when the observed process possesses certain properties, for example, the Chapman-Kolmogorov condition, the Markov property, stationarity and ergodicity are hold. The above properties of the process determine the evolution equations for the transition probability in the form of the Fokker-Planck equations already mentioned or Hamilton-Jacobi type equations, Schrödinger equations, and so on.

Topological prohibitions, implying the existence of such prohibition by physical mechanisms, determine other scenarios for the evolution of trajectories, that is, the probability of transition from one vector to another for a fixed number of steps. Moreover, in the interpretations of the process as a random walk on a lattice or continuum, processes are realized with allowance for the prohibitions imposed by the nontriviality of the homotopy types of the degeneracy space. Evolution equations at the same time are complicated. And to obtain evolution equations, it is required to introduce three-point, four-point, etc. density of the distribution function for transition probabilities. In the present, brief review of approaches should be mentioned often; there are cases where the probability density function of the process or the transition probabilities are not Gaussian but have a so-called heavy tail in its distribution, expressed as

$$P(R, L) \sim \frac{L}{\|R\|^{\alpha+1}}. \tag{19}$$

Thus, the approaches described above lead to the evolution equations with fractional derivative [25].

5. Construction of automata

This section is devoted to the description of a family of automata, analogs of Turing machines, allowing to formalize and, ultimately, take into account the above difficulties in the development of algorithms of reliable prognosis and numerical estimates of RUL, without which it is sometimes impossible to build and optimize the maintenance strategy and also implement the management task when choosing a class of trajectories that optimize the operation modes of technical objects.

The transition from abstract models to the construction of recognizing and predictive automata occurs in two stages. By recognizing automata in this case, we mean a set of single-tape Turing machines. At the first stage, the state space described in the previous section and in more detail in [7, 22] is constructed. The second stage involves the construction of the symbolic space described in the work [18–21]. For this, the transition from the initial state to the final state on the state space is represented as the product of matrices of special form acting in the affine space:

$$\Omega_{i,k}\begin{pmatrix} n_1 \\ n_2 \\ \cdot \\ n_i \\ n_k \\ 1 \end{pmatrix} \stackrel{\text{def}}{=} \begin{pmatrix} 1 & 0 & 0 & 0 & 0 & 0 \\ 0 & 1 & 0 & 0 & 0 & 0 \\ \cdot & \cdot & \cdot & \cdot & \cdot & \cdot \\ 0 & 0 & 1 & 0 & 0 & 1 \\ 0 & 0 & 0 & 0 & 1 & -1 \\ 0 & 0 & 0 & 0 & 0 & 1 \end{pmatrix}\begin{pmatrix} n_1 \\ n_2 \\ \cdot \\ n_i \\ n_k \\ 1 \end{pmatrix} = \begin{pmatrix} n_1 \\ n_2 \\ \cdot \\ n_i + 1 \\ n_k - 1 \\ 1 \end{pmatrix} \tag{20}$$

The product of such matrices eventually transforms the initial state vector into the final state:

$$\boldsymbol{\Omega}_{N^*} \stackrel{\text{def}}{=} \left(\prod_{1}^{N*} \boldsymbol{\Omega}_{i,k}\right) \text{ for } \forall\,(i, k) \tag{21}$$

The symbolic space in this case is a space P whose dimension is equal to the number of columns of the frequency histogram of the vectors of the initial or final. For definiteness, the dimension is increased by adding columns with zero values, thereby encompassing the physically permissible range of values of the observed signal. The successive multiplication of matrices of elementary steps (coordinatewise) with subsequent renormalization determines transition probabilities in the form of a matrix $\widehat{w}_{k,m}$ defined by the operator Ω_N .

The equation describing the evolution of certain states of the Turing machine in this way is a known balance equation [5]:

$$\frac{\partial p_m}{\partial t} = \sum_k \left[\widehat{w}_{k,m} p_k - \widehat{w}_{m,k} p_m\right] \tag{22}$$

or the basic kinetic equation [4].

The renormalization, which determines the transition from the frequency representation to the probabilistic one, also allows us to interpret the change of states as a walk in an N^*—dimensional simplex Σ^{N^*}. Since each one-tap automaton processes only one cascade of wavelet coefficients, then for the entire set of cascades, the number of automata is equal to the number of wavelet coefficients of the signal in one period. Since in the general case there is a scatter of periods over lengths, the set of recognizing automata is determined for each interval in the histogram of the lengths of the periods, depending on the values of the lengths of the elementary intervals and the probable transition to packet wavelet decomposition. At a signal sampling frequency of 2 kHz, the number of automata is estimated from below by a number equal to 40×10^3. If vibration signals are analyzed in a wide frequency range, then the number of automata is estimated by orders 10^6.

6. Accounting for topological dynamics and how the automata work

Thus, the operation of an automaton reduces to change in its internal state when it shifts by one step in the input cascade of wavelet coefficients of the observed signal. It is assumed here that the equations describing the changes in the state of the automaton are independent of time, that is, the internal states of the automaton are stationary. In practice, small deviations from stationarity in the Levy metric are allowed [26]. The value of the permissible deviations is determined on the basis of the chronological database. Thus, the quasi-stationary nature is verified by checking the approximate fulfillment of the stationarity conditions of the basic kinetic equation Eq. (22):

$$\widehat{w}_{k,m} = \sum_m \widehat{w}_{m,k} \tag{23}$$

If the quasi-stationary conditions are violated, for example, a trend appeared in one of the columns of the histogram or in several columns, the prognosis of the evolution of the internal states of the automaton is determined already by solving the nonstationary basic kinetic equation. Meanwhile, under the quasi-stationary conditions, the change in the internal states of the automaton is possible. As an example, we can mention the noise-induced transitions [16]. In this case, the change in the type of internal states is connected with a slow evolution of the coefficients of polynomial approximation of stationary solutions of the basic equation [3, 20]. Thus, in the transition from the initial state vector in the state space of in the symbol space introduced above, there are many transition paths in the formalism of birth-death process. However, all paths under quasi-stationary conditions reduce to permutations in the commutative subgroup of matrices of elementary transitions $\Omega_{i,k}$. That is, the transition from one vector to another takes place under the condition that the form of the symbolic histogram is stationary or that small deviations in the Levy metric are assumed. Moreover, the set of transition paths are in the same homotopy class of k-fold loops of the degeneracy space. The scenario described is valid for Gaussian processes with zero correlation length over the time variable. As noted in this case and taking into account the symmetry with respect to time reversal, the degeneracy space is an m-dimensional projective space. When time correlations of nonzero length appear, the degeneracy of the group Z_2 occurred. When a trend appears in the signal, degeneration by subgroups of the permutation group is removed. The admissible class of paths, on which the transition from one state vector to another takes place, is narrowed. Further dynamics of the internal states of the automaton is associated with the creation of new filling cells and the destruction of some old ones. At the same time, new ways of transition from one state to another are determined. And with the appearance of new ways, the estimate of the time to reach the class boundary or the field of the failure is changed. A complete reconstruction of the whole class of paths occurs when the homotopy type of the degeneracy space is reconstructed, for example, under noise-induced transitions reminding the second-order phase transitions in condensed media.

7. Reproduction and birth of automata with increasing complexity

However, the constructed set of automata is still not enough for an effective prognosis. This deficiency is closely related to topological dynamics, in particular, to the already mentioned effects of excluded volume. In other words, taking into account the homotopy classes of paths, that is, in those important cases, when the processes under consideration are not Markovian, the Chapman-Kolmogorov identity is not satisfied. In the analytic approach, three-particle and then many-particle distribution functions are considered in such cases. Here, the system of equations for these functions can have infinite dimension. Another analog of such equations is the transition from evolution equations in PDF to equations in moments, where an infinite-dimensional system of equations also appears. Most often, such equations are solved by truncating in dimension, assuming that the finite-dimensional part of the system of differential equations approximates an infinite chain of equations in some sense. In the case when multiparticle distribution functions are introduced to obtain solutions, it is assumed that, beginning with a certain number, many-particle functions are assumed to be approximately Markovian, that is, are represented as a product of multipoint functions of lower orders.

The examples given represent some analogies for completing the construction of a set of automata. The family of constructed automata with the necessity for complete account of topological dynamics must be supplemented by some additional properties. The analogies described above demonstrate what properties a family of automata should possess.

Additional properties of the family of automata:

1. Automata must be pairwise interacting.

2. In a number of cases, automata must analyze the described situations, that is, construct automata by merging one-tape automata, thereby passing to automata with increasing complexity.

A pair interaction between automata can be introduced in different ways, depending on the language describing these automata in terms of evolution equations or in constructing the Feynman path integral. In this case, the simple way of constructing interacting automata is to introduce interaction through a statistical interaction, taking into account this interaction from the first principles, when the measure of interaction is $d(P,P')$ [11, 26]:

$$d(P,P') = \left\{ \int_{\Omega} |\psi(\omega) - \psi'(\omega)|^2 dQ(\omega) \right\}^{1/2} \tag{24}$$

where P, P' are two probability measures:

$$Q = \frac{1}{2}(P + P') \tag{25}$$

$$\psi = \left(\frac{dP}{dQ}\right)^{1/2} \psi = \left(\frac{dP'}{dQ}\right)^{1/2} \tag{26}$$

is the Radon-Nicodym density.

There are more complex forms of interaction by introducing a potential or some vector field, as described in the work [6]. With the pair interaction between automata taken into account, it becomes possible to construct a graph whose automata are placed in 0-dimensional vertices. Since we are talking about automata on the wavelet cascades, the resulting graph allows us to analyze the situation with a root cause prognostics if the interaction is defined for automata with different time indices. Or, if automata interact with different scaling indices, then an analysis of multiscale processes or processes that occur at different scale levels and are interconnected becomes available.

8. Conclusions

The result of the completed constructions is the family of predictive automata. The set of automata is large, and depending on the sampling frequencies of incoming signals from sensors installed on the mechanisms, it is estimated from 10^6 to 10^9. The automata themselves represent some analog of the Turing machine. In this case, the set of automata interacts in pairs. The interaction leads to the construction of more complicated automata and is an analogy of transitions to many-particle distribution functions or their densities. Thus, the family of interacting automata generates the next generation of automata with increased complexity. Many features of the operation of automata and methods for constructing state spaces and degeneration spaces remain outside the scope of this chapter. It is only necessary to note that as the state space it is considered a sequence nested with respect to the dimension of spaces, as is the sequence of path spaces whose multiplicity can tend to infinity. In such limiting cases, infinite-dimensional symmetry groups appear. And when implementing limit transitions, there are mathematical models that allow us to algorithmize the problem in some sense.

The complexity of family of automata is determined by the complexity of topological dynamics. If we are talking about the observed signals, then the account of symmetry groups, the appearance, and the removal of degeneracy in different subgroups are determined by the complexity of the signals themselves, reflecting in turn the complexity of physicochemical processes occurring in complex mechanical and electronic systems at various scale levels. Symmetry groups appear in all existing time series. Most often the groups of symmetries of incoming signals are caused by the concrete physical processes of the failure physics occurring in complex mechanisms in the presence of friction, gas hydrodynamics, physical and chemical processes, etc. In the overwhelming number of cases, the physical models of the failure physics confirm this fact. Turning to concrete implementations of the family of prognostic automata, it should be noted that, in spite of the complexity of topological dynamics, specific algorithms prescribe fully realizable requirements for the costs of supporting their work in parallel architectures. In this case, as practice has shown in the operation of automata, their structures are optimized.

In addition, in the optimization process, there are opportunities that allow some automata to be transferred to the computing power of onboard computers when it comes to, for example, monitoring of mobile objects, in particular, monitoring of all kind of transport. At the same time, onboard automata perform not only signaling functions but also are able to manage remote computing cores, thereby optimizing the computational processes on the remote computing cluster. And to the contrary, the automata of computing cluster can change the structure and functionality of peripheral automata, located on onboard computers or other computational capacities inherent in microcontrollers of onboard electronics. Returning to the complexity of topological dynamics, it should be noted that the automata for prognosis that support this complexity allow us to formalize and algorithmize the models of the root cause prognosis and, in the end, algorithmize the tasks for the intellectual self-maintenance and self-recovery systems.

From the model constructed above, a certain hierarchy of prognostic problems also follows, since the set of physically acceptable trajectories is divided into the classes of homotopic equivalence. In turn, the classes themselves are changed during transformations of the space of degeneracy; other admissible trajectories and their classes appear that differ from the previous ones and have their own predictors and time estimates of the RUL. This means that with each transformation of the space of degeneration there is a change in the prognosis and changes in the RUL estimates. Thus, with RUL estimates, it is necessary to take into account not only the time to reach the class boundary but also the time to reach the moment when the transformation of degeneration space begins, for example, the time to reach the bifurcation set in the bifurcation tasks of the stationary solution of the Fokker-Planck equation. Another example is the time to reach a certain critical value of the amplitude of multiplicative noise. And each time the task of determining the RUL is updated. A good example of such an update is the calculation of the probability density of the transition from the original value to the preassigned one when determining the RUL in the representation of the probability density of the transition for a fixed time in the form of the Feynman path integral. At the same time, a change in the class of admissible transition trajectories, when the degeneracy by one of the isotropy groups or by its subgroups is removed, changes the evolution equation itself. In some cases, with the emergence of new topological obstacles, the Smoluchowski-Chapman-Kolmogorov identity does not hold the system and becomes non-Markov etc.; finally, an integro-differential equation appears as the Fokker-Planck evolution equation. In this case, all the previous predictors are changed, as well as all the time estimates. And so it happens with each new transformation of the space of degeneration.

The family of interacting automata presented here changes traditional approaches to the learning of automata in recognizing early predictors of failure, in other words, in identifying the characteristics of the trajectories, the movement along which leads to the boundaries of the failure regions. In the examples described above and from the general model, it should be that learning is reduced to a set of segments of wavelet coefficient cascades as long as the automaton output to the quasi-stationary regime is not going to happen. For simple automata, the segment length is estimated at about 1000 full cycles of the engine operation or the number of revolutions of the turbine shaft. Further, the algorithm during monitoring verifies compliance with the conditions of quasi-stationarity. At the birth of more complex automata with an increase in their dimension, each added dimension increases the length of the segments.

And in conclusion, it is necessary to say a few words about the set of early predictors of failure. In accordance with the hierarchical construction of the prognostic model, when removing the next degeneration by one of the isotropy subgroups in each new class of trajectories, their predictors are determined. That is, the inheritance of predictors in the transition of the trajectory from one class to another is not necessary. And this non-obligation is connected with the mapping of the gopotopic groups of the previous space of degeneration into the space of degeneration after its transformation. Part of the predictors may persist, another part may disappear, and new predictors may appear. The listed mutations of the set of predictors are determined by the specific structure of the degeneration space, that is, by the set of symmetry groups and isotropy groups. In this case, the trajectory itself or its characteristics, for example, configurational entropy, can act as a predictor, along with other types of the Kullback type of entropy.

Another type of predictor exists, and here again the analogy between the topological singularities of condensed media and the singularities of multiple loop spaces is appropriate. We are talking about the structure of the singularity core. Conducting the noted analogy, if the trajectory passes through the core of the singularity, then the effect of changing the permissible number of trajectories gives rise to changes, for example, of pointwise holder regularity of the trajectory.

In terms of the evolution of the internal states of a set of interacting automata, the above conclusions are expressed as additional conditions imposed on the densities of the distribution functions and transition probabilities for automata in any dimension.

This chapter is mainly devoted to the presentation of theoretical prognostic models and the basic ideas of constructing predictive automata. Demonstration of examples of the work of predictive automata and more detailed description of the predictors of the early prognosis will be continued in the next edition of IntechOpen book *Prognostics* edited by Prof. Fausto Pedro García Márquez.

Author details

Aleksandr Kirillov[1], Sergey Kirillov[1]*, Vitaliy Iakimkin[1] and Michael Pecht[2]

1 SmartSys Prognosis Center, Moscow, Russia

2 Center for Advanced Life Cycle Engineering, University of Maryland, Maryland, USA

*Address all correspondence to: skirillovru@gmail.com

References

[1] Available from: http://phmconf.org/phmtopics.html

[2] Kirillov S, Kirillov A, Iakimkin V, Kaganovich Y, Pecht M. Systems of preventive cardiological monitoring: Models, algorithms, first results, perspectives. In: Farhadi H, editor. Medical Internet of Things (m-IoT): Enabling Technologies and Emerging Applications. London: InTechOpen; in press; 2018. ISBN: 978-953-51-6138-7

[3] Kirillov S, Kirillov A, Iakimkin V, Khodos F, Kaganovich Y, Pecht M. PHM applications in medicine and medical implantable device. In: Proceedings of the IEEE Prognostics and System Health Management Conference (PHM-Chengdu). 2016. DOI: 10.1109/PHM.2016.7819755

[4] Nicolis G, Prigogine I. Self-Organization in Nonequilibrium Systems: From Dissipative Structures to Order through Fluctuations. Hoboken, New Jersey: A Wiley-Interscience Publication, Wiley; 1977

[5] Van Kampen NG. Stochastic Process in Physics and Chemistry. Amsterdam: North Holland Publishing Company; 2007

[6] Ziman J. Model of Disorder. Cambridge: Cambridge University Press; 1979

[7] Khodos A, Kirillov A, Kirillov S. RUL assessment and construction of maintenance strategies for engineering objects. In: Proceedings of the 8th World Congress on Engineering Asset Management (WCEAM 2013) & the 3rd International Conference on Utility Management & Safety (ICUMAS). 2013. pp. 1497-1508. DOI: 10.1007/978-3-319-09507-3_127

[8] Kobayashi S, Nomizu K. Foundations of Differential Geometry. Vol. 2. Wiley Classics Library; 1996. ISBN 0-470-49648-7

[9] Adams J. Infinite Loop Spaces. Princeton: Princeton University Press; 1978

[10] Mermin ND. Topologies theory of defects. Reviews of Modern Physics. 1979;**51**(3):591-648

[11] Hida T. Brownian Motion. New York: Springer-Verlag; 2012

[12] De Gennes PG. Scaling Concepts in Polymer Physics. New York: Cornell University Press; 1979

[13] Fuchs D, Fomenko A, Gutenmacher V. Homotopic Topology. Budapest: Akadémiai Kiadó; 1986

[14] Dubrovin B, Novikov S, Fomenko A. Modern Geometry: Methods and Applications. 2nd ed. Berlin Heibelberg New York: Springer-Verlag; 1991

[15] Toda H. Composition Methods in the Homotopy Groups of Spheres. Princeton: Princeton University Press; 2016

[16] Horsthemke W, Lefever R. Noise-Induced Transitions Theory and Applications in Physics, Chemistry, and Biology. Berlin Heibelberg New York: Springer-Verlag; 2007

[17] Dremin I, Ivanov O, Nechitailo V, Furletov V, Terziev V. Precursors of stall and surge processes in gas turbines revealed by wavelet analysis. Control Engineering Practice. 2002;**10**:599-604. DOI: 10.1016/S0967-0661(02)00005-9

[18] Kirillov A, Kirillov S, Pecht M. The problem of PHM cloud cluster in the context of development of self-

maintenance and self-recovery engineering systems. In: Proceedings of the 8th World Congress on Engineering Asset Management (WCEAM 2013) & the 3rd International Conference on Utility Management & Safety (ICUMAS); 2013. pp.1509-1520. DOI 10.1007/978-3-319-09507-3_128

[19] Khodos A, Kirilllov A, Kirillov S. Multisolution analysis time series data and RUL estimate. In: Proceedings of the 4th IEEE Conference on Prognostics and System Health Management (PHM). Vol. 33. 2013. DOI: 10.3303/CET1333057

[20] Kirillov A, Kirillov S, Pecht M. Remote calculating PHM cluster: The first results. In: Proceedings of the 4th IEEE Conference on Prognostics and System Health Management (PHM). Vol. 33. 2013. DOI: 10.3303/CET1333045

[21] Iakimkin V, Kirillov A, Kirillov S. Preprocessing of the observed data and the recognition of the hidden signs of fault. In: Proceedings of the 4th IEEE Conference on Prognostics and System Health Management (PHM). Vol. 33. 2013. DOI: 10.3303/CET1333012

[22] Kirillov A, Kirillov S, Pecht M. The calculating PHM cluster: CH&P mathematical models and algorithms of early prognosis of failure. In: Proceedings of the Prognostics and System Health Management Conference (PHM- Beijing). 2012. DOI: 10.1109/PHM.2012.6228771

[23] Iakimkin V, Kirillov A, Kirillov S. Distributed pre-processing of telemetry for mobile engineering objects. In: Proceedings of the 8th World Congress on Engineering Asset Management (WCEAM 2013) & the 3rd International Conference on Utility Management & Safety (ICUMAS). pp. 463-472. DOI: 10.1007/978-3-319-09507-3_40

[24] Kleinert H. Path Integrals in Quantum Mechanics, Statistics, Polymer Physics, and Financial Markets. Singapore: World Scientific Publishing Co; 2004

[25] Zaslavsky G. Chaos, fractional kinetics, and anomalous transport. Physics Reports. 2002;**371**(6):461-580

[26] Feller W. An Introduction to Probability Theory and its Applications. Vol. 2. Hoboken, New Jersey: John Wiley & Sons Inc; 1991

Chapter 3

Decision Maintenance Management Problems in Agriculture Engineering by Constructive Geometric Modeling Methods

Tojiddin Juraev

Abstract

Extension of functional possibilities of tools is one of the main ways to increase the maintenance property of technical means. It especially actually in modern agricultural production, based on precision agriculture technologies, where using technical means must provide: reduction of cost, conservation of ground fertility, saving energy-resources, improvement labor conditions, and increase machines capacity. One of efficient way to solve these problems is using geometric modeling methods and systems in designing technical means. Geometric modeling, as one of the varieties of the synthetic methods of design, is a theoretical base for different technologies of these methods, like industrial design and CAD technologies. In this chapter, as examples, the following case studies will be considered: development of multifunctional mold board by geometric modeling, for increasing its maintenance property; integration role of CAD technologies in PLM, including in maintenance management; and visualization of production design process of technical means according to maintenance criterions.

Keywords: geometric modeling, production design, CAD technologies, product life cycle, multifunctionality, mold board's surface, maintenance criterions and properties

1. Introduction

Extension of functional possibilities of tools is one of the main ways to increase the maintenance property of technical means. It particularly actually in modern agricultural production, based on precision agriculture technologies, where using technical means must provide: reduction of cost, conservation of ground fertility, saving energy-resources, improvement labor conditions, and increase machines capacity. One efficient way to solve these problems is using geometric modeling in designing technical means. Geometric modeling, as one of the varieties of the synthetic methods of design, is a theoretical base for different technologies of these methods, like production design and CAD technologies. In this chapter, as examples, the following case studies will be considered: development of multifunctional mold board by geometric modeling, for increasing its maintenance property; integration role of CAD technologies in PLM, including in maintenance management; and visualization of production design process of technical means according to maintenance criterions. In recent years, increased variety of applicable machines and technologies has come to exist in the world in all spheres of human activity, especially in planning their functional possibilities. So development of tools that increase their functional possibilities is one of the most important problems of modern engineering and design activity. Expanding the functional possibilities of these tools will not only increase their capacity but also reduce specific consumptions of materials. These aspects are actually in creation resource and energy saving technical facilities, that is, main engineering activity, key direction in which is considered production design. This problem is considered in the same way actual and in condition of the strategic development of the Republic of Uzbekistan [1]. The solution to these problems is directly connected with the geometric modeling, which is based on the modern problems of the production design [2–4].

2. Development of bulldozer's multifunctional mold board by geometric modeling, for increasing its maintenance property

2.1 Designing the types of mold board's working surface

We shall consider the problem in moldboard-type tools as an example. It is well known that mold boards, as the main tool in plows, bulldozers, graders, and other specific machines, are intended for performing the preparing works in agriculture and melioration, ground works in road construction and engineering preparation of territory, as well as in municipal sphere and etc. In the classical variant bulldozer, the mold board is a frontally located cylindrical working surface, which moves earth or other mass, prism lug of the ground in the required direction and amount [5, 6]. For expansion functional possibilities of mold board, there are also development in different constructive variant, with changing location working organ and different working surface (**Table 1**). But these developments are basically directed at the expansion of their maintenance (functional) possibilities, for executing work of certain nature [5, 7]. The solution to these problems is directly connected to the geometric modeling, which is based on modern problems of the industrial design [2–4]. The result of the using the production design at development of mold board type tools on base of constructive geometric modeling is a "design-development" mold board, which possible produce in three types of working surface design. We shall consider the design-development working surface of mold board consists of pieces of surface. For the base of the models, we take multifunction surface consisting of linear surfaces, which are broadly used for designing mold boards (**Table 2**).

The design-development to construct a geometric model of a mold board's working surface applicable for work execution of the different nature raises: technical, technological and economical factors of the designed technology, allows more flexible control its functional possibility, solving constructive problems [2, 8, 9].

The analysis of existing mold board designs and studies on their improvement shows that creating a new design that can increase their functional possibilities, using constructive geometric modeling method, has a broad prospect [5, 7, 8, 10, 11]. The constructive geometric design of mold board's working surfaces can possibly be divided into three types: (1) design consisting of unbroken surface (**Figure 1**); (2) design consisting of surface pieces (the sections) (**Figure 2**); and (3) design consisting of surface elements (the plates) (**Figure 3**). Herewith possible creating away the prospects of the primary using these design on example:

DOI: http://dx.doi.org/10.5772/intechopen.81969

No	Mold board construction	Surface type	№	Mold board construction	Surface type
1		Cylindrical (Poland)	4		Combined (Czech)
2		Conidial (Finland))	5		Cylindroid (Sweden)
3		Conical (USA)	6		Planar (France)

Table 1.
Using linear surfaces in bulldozers' mold board types.

No	Geometry of surface	Using in tools
1.	Frontal planar surface	Moldboard of channel defogger
2.	Inclined planar surface	Bush cutting mold board of bulldozer
3.	Frontal cylindrical surface	Frontal mold board of bulldozer
4.	Inclined cylindrical surface	Bucket mold board of scraper
5.	Frontal conical surface	Moldboard of grader
6.	Inclined conical surface	Frontal plow's mold board
7.	Cylindroid surface	Universal plow's mold board
8.	Conidial surface	High-speed plow's mold board
9.	Hyperbolic-parabolic surface	Hyperbolic body plow's mold board
10.	Helicoid surface	Helicoid body plow's mold board
11.	Torsos surface	Cultural plow's mold board
12.	Combined surface	Combined body plow's mold board

Table 2.
Using linear surfaces in mold board-type tools.

(1) unbroken design for production of polymeric mold boards; (2) sectional design for expansion of the functional possibilities and increasing the ease of manufacturing production mold boards; and (3) plate design for the best management production and functional, working, and other quality mold boards. Developmening the working surface of mold boards that can be applied in different industries needs to consider the technical, economical, and technological factors of the designed machines. So, design-development of constructive geometric model of mold board's working surface, though exist in the designs of the considered machines, will allow more flexible control of the functional possibility of the mold board and solve the above delivered constructive problem [2, 8, 9].

2.2 Geometrical modeling of mold board's transformed surfaces

There is giving formative line—l of cylindrical surface—Φ horizontally, and P—plane of directory curve—m dispose perpendicular to these formatives on medium them. This plane crossing with working surface—Φ is divided into two equal parts, Φ_a and Φ_b, simultaneously being the symmetrical plane to these working surfaces. We shall choose line k on symmetrical plane, which will possibly conduct the bunch of the planes. These planes crossing with working surfaces Φ_a and Φ_b form curves of intersection. We shall mark these planes on both sides of the symmetrical plane P accordingly $P_1, P_2, ..., P_n$ and $P_1', P_2', ..., P_n'$, as well as curves of the intersection on working surfaces Φ_a and Φ_b accordingly $m_1, m_2, ..., m_n$ and $m_1', m_2', ..., m_n'$. At

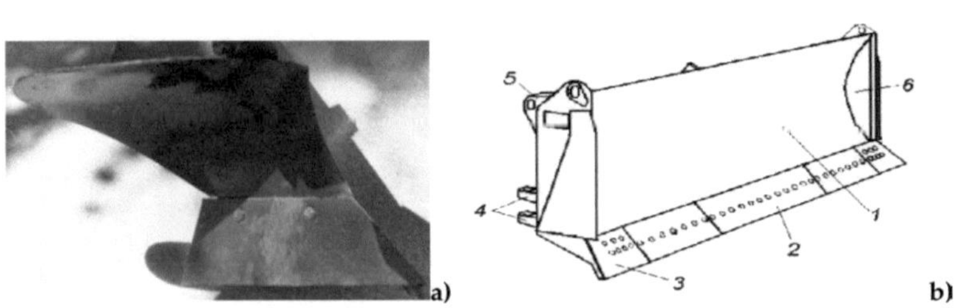

Figure 1.
Traditional construction of plow (a) and bulldozer (b) mold board.

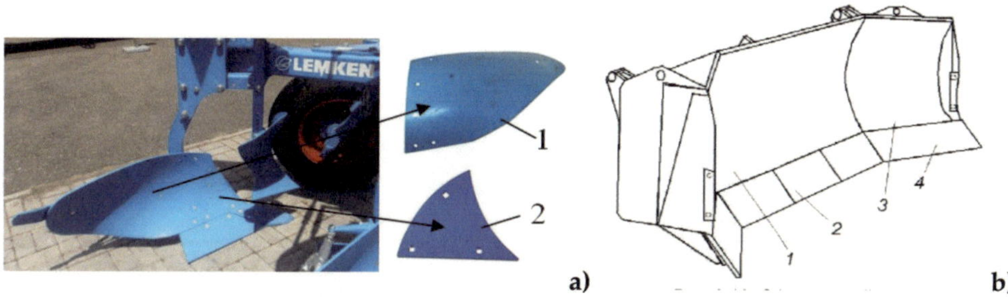

Figure 2.
Sectional construction of plow (a) and bulldozer (b) mold board: "1—wing" and "2—breast" of plow's body; "3—frontal" and "4—side" sections of spherical mold board of bulldozer.

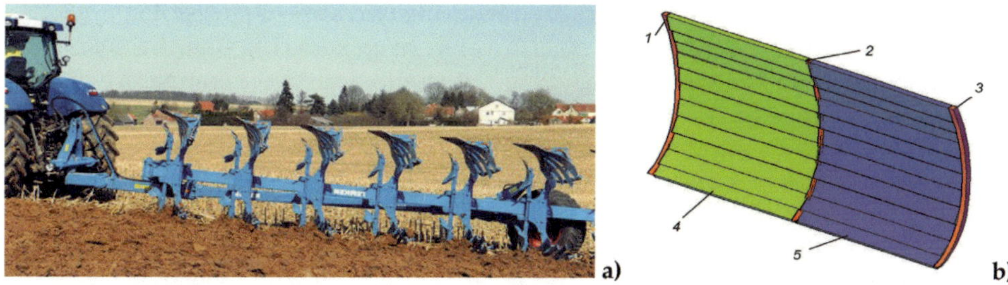

Figure 3.
Plate construction of plow (a) and bulldozer (b) mold board: 1—right, 2—middle and 3—left guiding frames, 4—right and 5—left formative plates (*construction offered by author).*

angles between planes and symmetrical plane—$\boldsymbol{P}$, we shall accordingly mark $\boldsymbol{\alpha_1}, \boldsymbol{\alpha_2}, \ldots, \boldsymbol{\alpha_n}$. Each pair of surface intersection curve $\boldsymbol{m_1}, \boldsymbol{m_1}'; \boldsymbol{m_2}, \boldsymbol{m_2}'; \ldots; \boldsymbol{m_n}, \boldsymbol{m_n}'$ are formed accordingly by pair of planes $\boldsymbol{P_1}, \boldsymbol{P_1}'; \boldsymbol{P_2}, \boldsymbol{P_2}'; \ldots; \boldsymbol{P_n}, \boldsymbol{P_n}'$, are symmetrical, where $\boldsymbol{k}$ is an axis of the mirror image pair of curves on working surfaces $\boldsymbol{\Phi_a}$ and $\boldsymbol{\Phi_b}$ (**Figure 4a**). So at rotation pair planes $\boldsymbol{P_i}$ and $\boldsymbol{P_i}'$ with surfaces $\boldsymbol{\Phi_a}$ and $\boldsymbol{\Phi_b}$ around axis $\boldsymbol{k}$ corresponding to angle $\boldsymbol{\alpha_i}$, planes $\boldsymbol{P_i}$ and $\boldsymbol{P_i}'$, as well as curves $\boldsymbol{m_i}$ and $\boldsymbol{m_i}'$ belonging to them, are combined. As a result of this operation, will be formed a rib on working surface, which separates the working surface into two halves. On the basis of this model, different constructive variants of the transformed mold board can possibly be developed, allowing the conversion from one working surfaces to another.

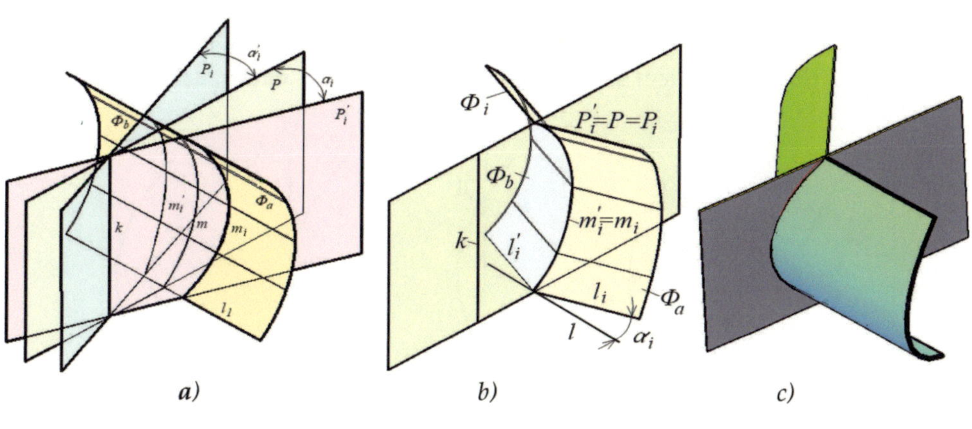

Figure 4.
Forming of moldboard's working surface with bilateral action: a) transforming sheme; b) moving ground mass to the sides; c) moving ground from the sides.

It is known that when designing the complexity technical forms, considered surface mentally differs on "geometric" and "working" surfaces and from one surface possible to get different working surfaces [4, 7]. So by means of the proposed model, as a result of rotation working surfaces Φ_a and Φ_b around axis $\boldsymbol{k}$ to angle α_i, a new working surface Φ_i is formed. Though given Φ and newly formed Φ_i cylindrical surfaces, they have a different working surface with different functional quality, where α enters as controlling parameter in the formation of Φ_i. Unlike the given surfaces Φ, a new working surface Φ_i promotes the improvement of directing actions of the moveable mass to the sides (**Figure 4b**) and from the sides (**Figure 4c**).

2.3 Giving the rotation axis of mold board's working surface

The process of the formation required working surface—Φ_i possible to control, except parameter α, as well as position of $\boldsymbol{k}$. In considering that the model rotation axis $\boldsymbol{k}$ is located vertically and has determined distance comparatively to Φ_i. However, change the position $\boldsymbol{k}$ greatly influences upon formation Φ_i. Here possible consider two parameters of $\boldsymbol{k}$: change the distance—$\boldsymbol{f}$, defined between fixed point $\boldsymbol{k}$ and $\boldsymbol{m}$, for instance base $\boldsymbol{k}$ and sock $\boldsymbol{m}$ on horizontal plane; as well as change of the slopping angle—β to horizontal plane. Under one and same angle α_i and the form of directory curve $\boldsymbol{m}_i$, change $\boldsymbol{f}$ will bring about change the mutual location pair of directory curves $\boldsymbol{m}_i$ and $\boldsymbol{m}_i'$ that will bring and to change constructive parameter of mold board with working surface—Φ_i. From considered by author, acceptable variants (**Figure 5**) for given problems are chose variants (*b*) a chord—AB and (*d*) a tangent in point—C, with the result that possible neglect the parameter f that simplifies the problem. Though the other variants too have such working surface, they can bring about complication in the constructive parameter of the mold board. However, when forming the surface Φ_i, in variant (*d*) rotation is produced in inverse direction than in variant (*b*). With the importance of the rotation angle α, we choose within $\mathbf{0} < \alpha < \alpha_{max}$, with the condition that planes $\boldsymbol{P}_i$ and $\boldsymbol{P}_i'$ must cross all forming surfaces Φ_i, where α_{max} is on $\boldsymbol{tg\alpha} = \mathbf{(1/2)/}\boldsymbol{b}$, and overhang of curve $\boldsymbol{b}$.

2.4 Parameters of designing working surface's directory curve

It is necessary to note the parameters on the form and position directory curve $\boldsymbol{m}$ of surface Φ. On condition of the problem form of directory curve—$\boldsymbol{m}$ is flat and fluent, with determined by curvature and concave side onward. Since these characteristic directory curves remain low-lying during the transformation of the surfaces, they shall select as topological parameters of curve, defining its

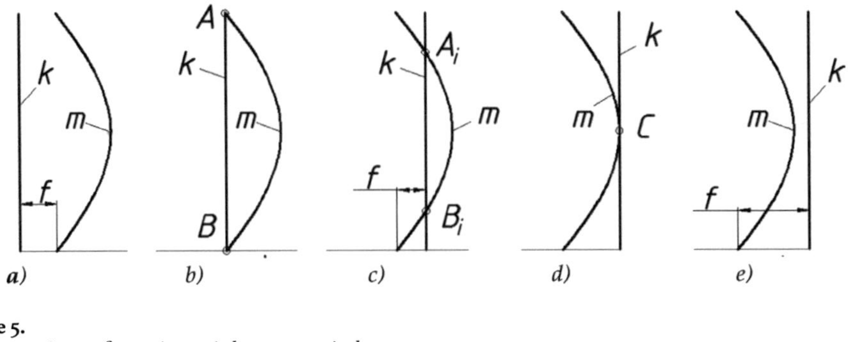

Figure 5.
Position variants of rotating axis k comparatively to m.

form. Consequently, such parameters of surfaces, as their type and curvature also remain low-lying and when forming the new surface Φ_i. The position of curve is assigned two parameters: overhang b and height h of curve. They shall be marked as constructive parameters, since they define the design of the mold board. The following variants possibly select the relative position constructive parameters of m, defined by typical point positions (**Figure 6**): lower (A) and upper (B) points define h, and extreme left and right (the pair from points A, B, C) points define b. These variants directory curves are possible to choose when designing the mold board depending on execution of its work. When changing f, in the vertical position of k, the dimension height of mold board h' in the same way remains low-lying. The parameter $\delta b_{max} = b_i - b$ derived after forming rib of surfaces Φ_i is situated opposite, for points, on which pass the rotation axis k (*right/left*—on bosom or *upper/lower*—on carrying).

2.5 Parameters of designing working surface's formatives

The criterions of the choice variant relative position of typical point of directory curve m on h and b, when designing mold board possible to explain, linking these points with typical positions of formatives l. For example, we shall select the following position formatives l, getting through typical points m on width b in respect to h, upper, lower, front, back, as well as average (on h or on b), and define their influence upon nature of the moving the moveable mass on working surface of the mold board (**Table 3**). From given table, it can be understood that the nature of the moveable mass on working surface is possible to control, having changed relations h and b, by changing the slopping angle β to axis k. Unlike vertical position, the slopping k on angle β onward or will back add the working surface except improvements of the shift of the moveable mass aside under its horizontal trimming (**Figure 7a**), as well as perfects the functional quality on shaping tilted lowering (**Figure 7b**) and ascent (**Figure 7c**) from moveable mass. This is the positions reached by change forming l, which present as well as plowshare, for horizontal plane on angle—φ, after forming Φ_i. The angle φ possible define by projection model, on base of descriptive geometry rules [12], using joining method (**Figure 7d**). Turning the horizontal plane on 90°, to joint it with frontal projection

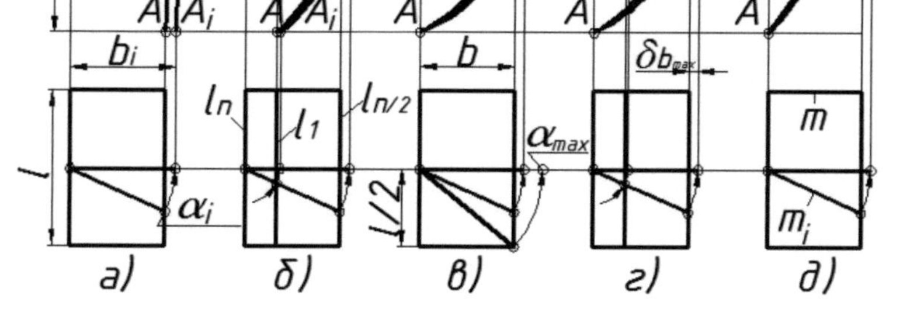

Figure 6.
Relative position variants of directory curve's constructive parameters.

No	On width *b*, in respect to *h* and through points			Nature of moving the moveable mass on worker of the surfaces
	Anterior	Average	Posterior	
1.	Superior—*B*	Not available	Interior—*A*	Powerfully postponed in before.
2.	Superior—*B*	Interior—*A*	Average—*C*	Partly is taken on breast and powerfully postponed in before.
3.	Superior/interior—*B*/*A*	Not available	Average—*C*	Completely taken on breast and powerfully postponed in before.
4.	Interior—*A*	Superior—*B*	Average—*C*	Completely taken on breast and weakly postponed in before.
5.	Interior—*A*	Not available	Superior—*B*	Completely taken on bosom.

Table 3.
Positions of formatives and their influence to working surface nature.

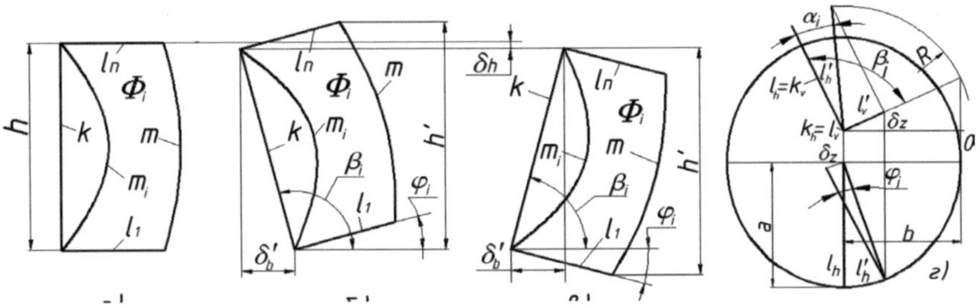

Figure 7.
Determination of inclined working surface geometric parameters.

combine the projections $\boldsymbol{k}$ and $\boldsymbol{l}$. Rotating $\boldsymbol{l}$ on angle αi, marked its $\boldsymbol{l}'$, easy find the frontal projection $\boldsymbol{l}_v'$. Since $\boldsymbol{l}$ revolves on frontal projection plane, perpendicular to $\boldsymbol{k}$, circle of the rotation $\boldsymbol{l}$ projects on the horizontal plane as an ellipse. By means of projection beams, find $\boldsymbol{l}_h'$ and define φ angle of the slopping $\boldsymbol{l}$ on horizontal plane using the square-wave triangle, also considered as the corner of the slopping of the plowshare. After transformation working surface $\boldsymbol{\Phi}$ on $\boldsymbol{\Phi}_i$ under inclined $\boldsymbol{k}$, will increase the height dimension $\boldsymbol{h}'$ of mold board though $\boldsymbol{h}$ decreasing on $\boldsymbol{\delta h}$. At, the higher part rib bends over onward or for lower part back, daring on distance $\boldsymbol{\delta b}'$. As a result of transformation, working surface changes the lengths corresponding to forming $\boldsymbol{l}_i$ within $\boldsymbol{0 < \delta b < \delta b_{max}}$, offset end forming belonging to rib to surfaces. In point, on which pass the rotation axis $\boldsymbol{k}$, length $\boldsymbol{l}_i$ is equal $\boldsymbol{\delta b = 0}$, but in nose (*upper* or *lower*) of a part it is equal $\boldsymbol{\delta b = \delta b_{max}}$.

2.6 Sections of designing working surface

The definable parameters got $\boldsymbol{\Phi}_i$ on two variants, and on positions of the descriptive geometry, make sure that $\boldsymbol{\alpha}_i$ parameters and $\boldsymbol{\Phi}_i$ are alike, but are mutually negative (**Figure 8a** and **b**) [12]. This allows to combine two variants in one design and as a result enlarges the functional possibilities of the designed mold board (**Figure 8c**). We can select five compartments of working surfaces on intersection lines. Alternate switching-on or switching-off of corresponding compartments will enable the mold board to work in three modes: moving the mass

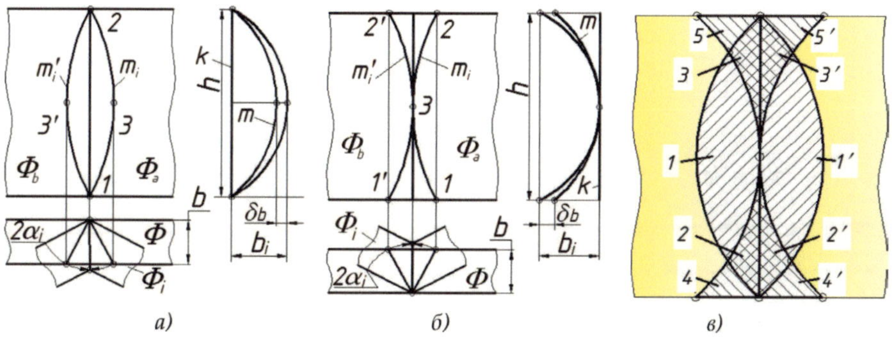

Figure 8.
Forming of working surface with bilateral action.

frontal, moving ground mass to the sides, or from the sides. The proposed device of geometric modeling-transformed working surface allows to develop a constructive geometric model of a multifunctional mold board. This development is intended for organizations to produce specific machines. Parameterization of mold board's working surface relieves designers' work, increases the choice a variant under development mold board's working surface, and allows effectively to solve the constructive problems.

3. Integration role of CAD technologies in PLM, including in maintenance management

The modern production is founded on using information science and communication technologies as CALS-technology (continuous acquisition and lifecycle support) or PLM-technology (information support of the product lifecycle management processes). PLM is an approach to designing and producing high-tech and scientifically based product, using information science and computer technology at all stages of the product life cycle [13].

This aspect actual in condition of developing countries, like Uzbekistan, where using these technologies is innovative process in production. One of the problems in this process is adapting them in production, that is, translating the engineering data to PLM system, by way of integrating PLM and CAD/CAE/CAM systems, using the product's engineering database at the base of PDM-technology (product data management).

The product's engineering data are possibly divided into three groups: structural (constructional), functional, and technological. Let me present to you the structural data, which we can call the geometric data, that are necessary for integrating CAD and PDM systems. The product's geometric data are used not only in enterprise where they are produced but at all stages of the product life cycle from designing to maintenance. So, creating the geometric database, using different forms of the geometric data (**Figure 9a**), is very important in the product life cycle.

As is well known, the product lifecycle includes the period from origin necessity for creating the product up to its liquidations in consequent exhaustion of consumer characteristic. Primary stages of product life cycle are selecting four main stages: designing, producing, maintaining, and utilizing.

Though life cycles of old and new products always form the unceasing cycle, because of brightly not images, traditionally life cycles of each product were considered separately, which during the initial stage was designing but finally cutting. However, author, founding on his conducting researches, offers to consider that

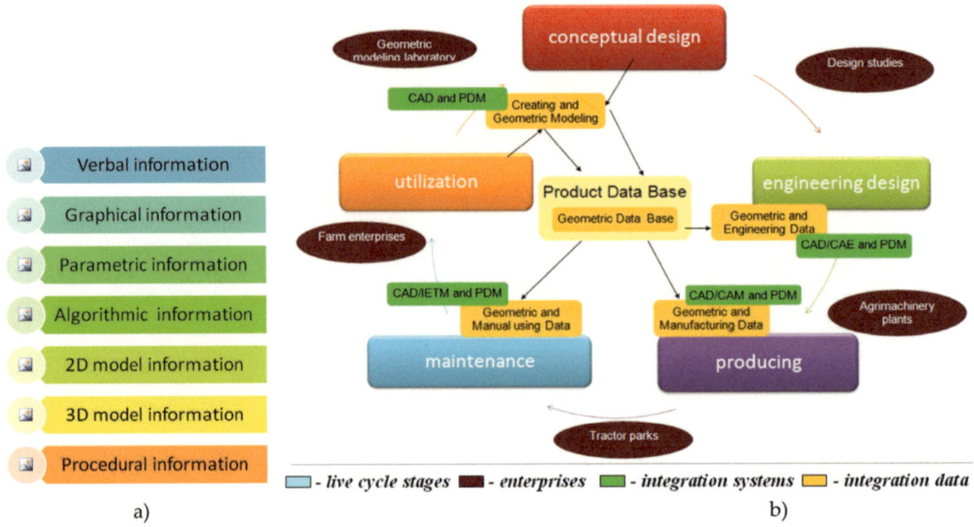

Figure 9.
Types and levels of geometric data (a). Integration role of geometrical modeling system in PLM (b).

that beginning of PLM is from creation an instrument from stone, bones and wood by primitive man. Since no one can reject that the base of modern industrial robot is an instrument of the stone age, the end of "old" product is a beginning of "new" product. It is possible to say that the beginning and the end of PLM connect with beginning and end of mankind on land. The present production conditions, in which production design steel play one of solving roles, relationship between "old" and "new" product lifecycle become reveals itself all more brighter. Coming from author's offers separate stage of the designing on two: conceptual and engineering design. Conceptual design stage is founded on the basis of geometric modeling and it is the closing stage of the product life cycle, having a causal relationship between "maintenance-utilizing" and "designing" stages. Nowadays, producing geometric modeling has become a primary method, facilitating designing. In this stage, the product will be designed on the basis of the relationships between exhaustion of consumer characteristic of old product and necessity for creating the new (innovative) product.

The need for geometric data arises at all stages of the product life cycle, particularly in the initial stage—"conceptual design" stage, when it is very high. Created at this stage, geometric database is directly or indirectly used also in the other stages of the product life cycle, by integrating CAD and PDM systems. It is necessary to note that the need to create a "new product" is basically formed in the maintenance of "old" product. Since it at this stage is not only used Geometric data of "old" product in maintenance but arises Geometric data of "new" product in designing.

As example, we shall consider creating the agricultural machinery tools' geometric database, which is necessary to enterprises that participate in these products life cycle [14]. Creating this database needs the review, classification, and analysis of appropriate information about agricultural machinery tools from a geometric standpoint. This will allow us to reveal general and individual geometric features of these tools that assist in an efficient management of product data for all participating enterprises in this process (**Figure 9b**). At present, the author is the leading researcher on development of theoretical bases and applying aspects of the geometric modeling of agricultural machinery tools. With the results from research, models, algorithms, and methods of designing these tools with mold board surface by geometric modeling were worked out.

	Influence of geometry on technical and technological characteristics	Considered constructions						Proposed design on advantage
		A	B	C	D	E	F	
1.	Influence on earth layer trajectory	0	+	0	–	0	0	B
2.	Influence on material quantity	+	+	–	0	+	–	BE
3.	Influence on process quality	0	+	0	–	0	0	B
4.	Influence on flat plugging	–	–	+	+	–	+	CDF
5.	Influence on multifunctionality	–	–	–	–	+	+	EF
6.	Influence on manufacturability	–	–	–	+	0	0	D
7.	Influence on complexity of construction	+	+	0	0	0	–	B
	Amount of disadvantages "–"		3	3	3	1	2	Advantages
	Amount of advantages "+"		4	1	2	2	2	
	Total score		1	–2	–1	1	0	
	Rating place		1	4	3	1	2	

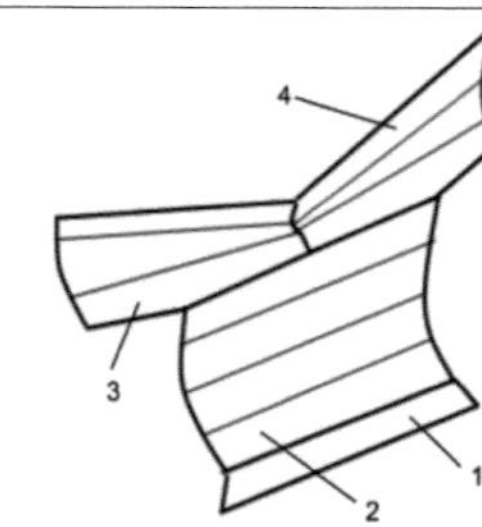

Table 4.
Screening of model selection process for designing mold board.

As a result, information sufficient for creating the geometric database in different forms was accumulated. We shall select the following forms of geometric data: verbal, graphical, parametric, algorithmic, 2D and 3D model, and methodical (that may include all geometric data forms). All elaborations are executed in *AutoCAD 2013* system.

4. Visualization of designing process of technical means according to maintenance criterion

The present pace of industry development requires the development and introduction in production innovation designing technologies. Using the methods of the production design gives the essential result in the process of developing technical object on different criterions engineers play one of key role in this, since saving to energy and resource, improvement to ease of manufacturing and functionality, as well as capacity, mainly depend on under development them technology and technical facilities [15]. Using of these methods require from constructors revealing the problems, which decisions are connected with geometrical modeling, on which is founded production design, as well as way of their decision. We shall consider using one of the methods of the production design—"*Concept selection,*" choosing the models to design. The moldboard have a complex technical form, long period of change on improvement their design, and have a universal geometric model. These factors allow using the production design in development mold board on geometric features, influencing on technician-technological features. As is well known, application in agricultural production plows has a different mold board constructions in accordance with their destination. The combination, on advantage of different criterions, considered constrictions in one new constriction, with necessary changes, by screening-method of "Concept selection" will allow to choose the models for development. The development conducting on the main types design moldboard plows, in which is taken into account row of the main criterion of the choice to models on geometric features (based on geometric data and parameters), presented by requirements of the producers and consumers (**Table 4**). As base, at estimations of the criterion, is chose design ***A***. Geometric features are valued upon their advantage (+) and defect (−). The features design, obviously not by specialists as advantage or defect, are conditionally evaluated neutral (0), coming from that considerations that they specifically do not influence upon these features. The visualization of the qualitative estimation and analysis of the features in such a way allow to choose the directions of modeling of designs on advantage, and then to combination of design. They are hereinafter offered development of the models, on basic model, occupied 1–3 places with provision for their advantage.

5. Conclusions

The proposed constructive geometric model of mold board's working surface allows to develop the multifunctional tools applicable in agricultural, engineering, road building, mining, and municipal service industries and in other branches of machinery use. Parameterization of mold board's working surface relieves designers' work, increases the variants choice under developing mold board's working surface, and allows effectively to solve the constructive problems. The integration of geometric modeling systems (CAD) and methods allows the efficient use not only in designing process and but also in production and maintenance processes of the technical facilities. Creating the product's geometric database by CAD technologies became one of the necessary tasks of production, particularly engineering products. In contemporary conditions of using CALS technologies, "conceptual design" stage of innovative product by methods and facilities of geometric modeling is the defining stage of the product life cycle. The visualization of the design-development process allows to develop a new product according to maintenance criterions. Efficient use of this method reduces design process time, labor, and material costs during the development of a new product.

Author details

Tojiddin Juraev
Descriptive Geometry and Engineering Drawing, Department of Bukhara Institute of Engineering Technology, Bukhara, Uzbekistan

*Address all correspondence to: tojiddin_1968@mail.ru

References

[1] Edict of the President of the Republic of Uzbekistan. About strategies actions on the further development of the Republic of Uzbekistan. Tashkent. Newspaper Xalq so'zi, No 28 (6692). 8 February 2017

[2] Production design. Principles, tools and techniques. ME 1007 Design principles. Concept selection. Available from: http://www.product_design_principles.pdf

[3] Tyalve E. Short Course of the Industrial Design. Translation Kunin PA. Moscow: Mashinostroenie; 1984. p. 192

[4] Juraev TX. Principles of Geometrical Modeling of Meliorative and Agriculture Machines Tools. Saarbrucken: Lambert Academic Publishing; 2015. ISBN 978-3-659-66832-6. Available from: http://dnb.d-nb.de

[5] Shestopalov KK. Machines for Earthworks: School-Book/Moscow institute of automobile road constriction (MADI). Moscow; 2011. p. 145

[6] Bosoy EC. and others. The theory, design and engineering of the agricultural machines. Moscow: Mashinostroenie. pp. 1978-568

[7] Mould boards. Available from: http://www.max-agro.ru

[8] Catalogue of special machines. ISP GROUP. Available from: www.isp-group.ru

[9] Voloshinov DV. The theory to automations of the designing object and processes on base of the methods of constructive geometric modeling [thesis]. Saint-Petersburg. 2010

[10] Biliev AA. The motivation of the technological scheme of the plow for smooth two-tier plowing [thesis]. Moscow: VIM; 1992. pp. 22

[11] Dubanov AA. Methods and algorithms to approximations of the technical surfaces turning round [thesis]. Moscow:MUPP; 1997

[12] Murodov ShK. and others. Descriptive Geometry. School-Book. Tashkent: IQTISOD-MOLIYA, 2008

[13] Jayakiran Reddya E, Venkatachalapathia N, Pandu Rangadub V. Development of an approach for Knowledge-Based System for CAD modeling. In: Materials Today: Proceedings; 5, 2018. pp. 13375-13382. Available from: https://doi.org/10.1016/j.matpr.2018.02.330/, www.materialstoday.com/proceedings

[14] Juraev TX. Creating the geometric database for product lifecycle management system in agricultural engineering. In: International Conference on Information Science and Communications Technologies ICISCT 2017 Applications, Trends and Opportunities. 2-4 November 2017; TUIT, Tashkent, Uzbekistan: IEEE Catalog Part Number: CFP17H74-CDR, ISBN: 978-1-5386-2167-7. Available from: https://www.researchgate.net/publication/321821311

[15] Juraev TX. Conceptual designing of Mould Board's surface by geometrical modeling. American Journal of Mechanics and Applications. 2017;5(4):28-33. DOI: 10.11648/j.ajma.20170504.11. ISSN: 2376-6115 (Print); ISSN: 2376-6131 (Online). Available from: http://www.ajmechanics.org/archive/621/6210504

Chapter 4

Maintenance Management with Application of Computational Intelligence Generating a Decision Support System for the Load Dispatch in Power Plants

Milton Fonseca Junior, Antônio Estanislau Sanches, Jandecy Cabral Leite, Jorge de Almeida Brito Júnior, Manoel Henrique Reis Nascimento, Carlos Alberto Oliveira de Freitas, Tirso Lorenzo Reyes Carvajal, David Barbosa de Alencar, Jorge Laureano Moya Rodriguez, Ubiratan Holanda Berzerra and Nadime Mustafa Moraes

Abstract

The development of a computational tool to support the decision of load dispatch according to the operational conditions of motors and generators of power plants is proposed, which are classified in relation to the probabilities of faults by a fuzzy system developed in this text, from indicators obtained from the analysis of lubricating oil, vibration analysis, and thermography of power generation equipment. The basis for the study is based on the principle of operation and operational conditions of the equipment to be dispatched for generation in a power plant, in addition to its particularities as specific consumption and the polluting emission for each equipment. In this way, this work aims not only to provide the tools to monitor these equipment but also, based on the management reports of vibration, temperature, and oil analysis, take corrective actions to maintain the necessary reliability and achieve the quality of the service through a preclearance procedure that takes into account the operating conditions of the equipment, obtaining performance indicators of the plan.

Keywords: economic load dispatch, management maintenance program, electric generators, power plants and fuzzy logic

1. Introduction

Most of the Brazilian thermoelectric park is completely shut down for months whenever the hydrological situation is favorable. As in the recent historical average hydroelectric generation has been 90% of its generation capacity for the system [1], idleness has prevailed in the thermal park since the plants can only be activated when the hydroelectric reservoirs are below 50% of its maximum volume. The contrast with the international reality is striking. In most countries, power plants with combined cycle of coal or gas typically do not experience inactivity during a long time period. Instead, they operate at the base of the system, being dispatched almost continuously. Additionally, thermals that in other electrical systems are used for generation of tip, with daily activation or at least in good part of the working days, such as open or thermal cycle gas engines with motors, in Brazil, can remain idle for long because they are not necessary in normal or favorable hydrology situations.

On the other hand, it is necessary to ensure the supply of electricity to consumers within standards of continuity and reliability. Besides, the lack of investments in the industry causes the loss of product quality, and the excess of investments can make the value of the product very expensive, disheartening its wear [2, 3].

To guarantee the quality and reliability of the electric power supply, it is necessary to perform an optimal load dispatch [2, 4]. Too many papers presented in the literature develop the load dispatch of the thermal plants considering that all the engines of the plant have good technical conditions, but this is not always true, so in this chapter, a method is presented for the pre-dispatch of load that takes into account the technical state of the plant's motors through diagnosis and making use of fuzzy logic.

The development of a computational tool to support the decision of cargo dispatch according to the operating conditions of the engines and generators of thermal plants is proposed, which are classified in relation to the probabilities of failure by a fuzzy system developed in this thesis, from indicators obtained from lubricant oil analysis, vibration analysis, and thermography of power generation equipment. The basis for the study is based on the principle of operation and operational conditions of the equipment to be dispatched for generation in a thermal plant, besides its particularities as specific consumption and the quality of pollutant sent by each equipment.

2. Maintenance systems and their application in thermoelectric plants

The ability of a generation source to meet an energy demand can be influenced by unexpected units of power-generating units. The tests were even more advanced to repair preventive maintenance measures but were not revised in the 1990s with maintenance and maintenance work on engines and generators.

In recent times, condition-based maintenance (CBM) has been introduced in industrial systems to preventively maintain the right equipment at the right time relative to its current "operating condition." The good state of operation of a generator can be represented mainly by conventional indicators such as oil temperature, harmonic data, vibration, etc. Then the importance of monitoring the motors/generators and their diagnosis for the dispatch of cargo to not have unexpected interruptions.

Most energy generation unit scheduling packages are considered preventive maintenance schedules for units over an operational planning period of 1 or 2 years in order to defray the total operation while meeting the requirements of system power and maintenance restrictions. This problem consists of verifying the generating units must be stopped from production. The generating unit should be regularly examined for safety. It is important to detect a failure in a power generation unit that can be used in the machines. The main indication is a suitcase case response. Therefore, the fixation and the key point are used in the proposed methodology. The issue is addressed as an optimization problem. The model is developed by determining the objective function, which is a net power reserve of the unit [5].

They point out that condition-based maintenance (CBM) is an approach that gathers and assesses information in real time, and based on this information, it recommends maintenance decisions based on the existing condition of the system. In the last decades, research on CBM has been rising rapidly due to the increment of computer-enabled monitoring technologies. It has been proved that CBM, when it is planned carefully, can improve the reliability of equipment reducing costs [6].

The factor of diagnostic importance (DIF) is frequently used for choosing preferences in maintenance activities at power supply sections of distribution systems. In [7] approach to assess a weighted cumulative diagnostic importance factor (WCDIF) for each section, which represents a good parameter for the ordering of maintenance activities, is developed. The methodology includes the effects of distributed generation (DG) and the loads. It was implemented as case studies in two distribution systems, so that, in the end, sorting lists of feed sections for maintenance activities were obtained [7].

In order to improve the reliability and efficiency of equipment, it is very important to apply the condition-based maintenance (CBM). A good maintenance activity has a close relationship with security and diminishes costs, making this issue even more attractive to researchers [8].

Proper maintenance can increase the company's productivity and increase its value in the market. The main study provided a robust model that can evaluate strategically important available technology and may exclude outdated and/or inappropriate technology. There are many researches in this field in which the number of models has been proposed, such as the maintenance management system, maintenance performance measurement, and maintenance performance indicators, but the details of the effectiveness of the predictive maintenance indicator specifically based on maintenance and conditions (MBC) with maintenance and management requirements using the analytical hierarchy (AHP) process are hardly available in the literature [9].

Basically, the process consists of monitoring parameters that characterize the state of operation of the equipment. The methods employed involve techniques and procedures for measuring, monitoring, and analyzing these parameters [10]. It can be related as oil analysis, ferrography, thermography, and vibration analysis.

Motor operation data in conjunction with vibration, oil, and temperature analysis data are collected periodically at the plant and are used in an integrated way to feed a fuzzy rule-based system, which returns the pre-dispatch scheduling of the plant for the period of interest, taking into account the state of operation of the machines (**Figure 1**).

Thermal power plants involve many mechanical and electrical systems that require constant analyzing of power production. The data obtained through this analysis are necessary for a good operation, maintenance, and evaluation of the performance of the plants. For this analysis the so-called distributed control systems (DCS) are often used. Nevertheless, the obsolescence of this equipment increases the risks of unavailability of the generating units, mainly in thermoelectric plants, where mechanical wear is elevated, due to the high temperatures and the chemical agents used for the production of electric energy [11].

Mean time between failures (MTBF) or mean period between failures is a value assigned to a particular equipment to describe its reliability. This value indicates when a device failure may occur. When this index is high, the reliability of the equipment and, consequently, the maintenance will be also evaluated as excellent.

The average (MTTR) time for repair is a measure on the basis of repairable item maintenance. It represents the average time required to repair a component failure or mathematically expressed equipment, that is, corrective maintenance.

Oil analysis: The initial purpose of oil analysis of a lubricated assembly or a hydraulic system is to economize by optimizing the intervals between the exchanges. As the analyses carried out resulted in indicators that report on the wear

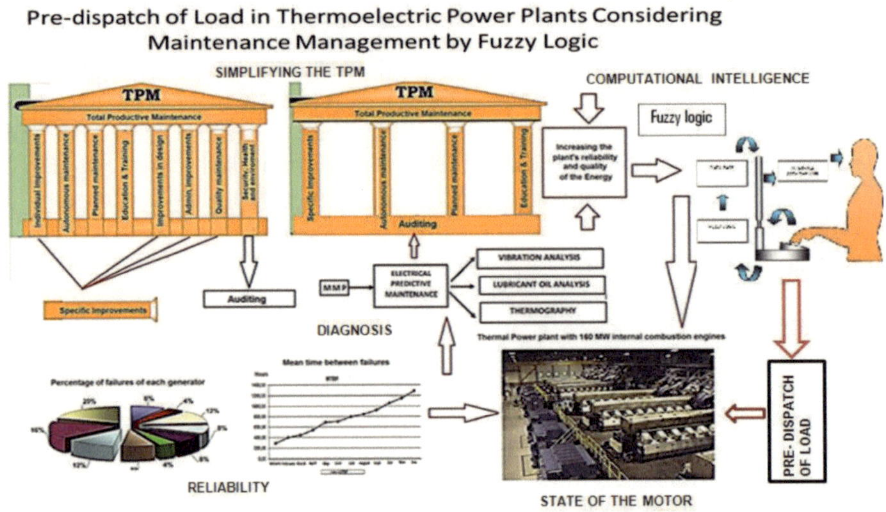

Figure 1.
Methodology. Source: Authors.

of the lubricated components, the second objective of this process became defect control for predictive maintenance [12].

- In the upper left, you can see the simplification of the eight pillars of the TPM, for four pillars.
- The left-center part shows the diagnostic activities that allow to know the technical state of the motors, to know whether or not they can be used in the pre-dispatch of load.
- In the lower left, we show the reliability analyses, which together with the diagnosis allow us to know when it is possible for each motor to fail to consider it in the pre-dispatch of load.
- The right part shows the application of the fuzzy logic, to perform the pre-dispatch of load, according to the fuzzy rules that meet the technical state of the engines.

3. Fragments used for implant and end methodology TPM program

This chapter presents a new solution proposal, which includes the pre-dispatch of load focused on the operational conditions of the machines using computational tools, specifically fuzzy logic. This application incorporates some novelties, such as good maintenance management through TPM program for decision-making, including performance indicators of the generating units such as vibration, lubricating oil, and temperature, analyzing if the generating unit will operate and maintain reliability or will get into maintenance due to poorly diagnosed performance.

3.1 Pillar of specific improvements (recommended group: coordinators of ME, MA (plant managers), MP, and SMA)

Objective: To maximize the overall efficiency of the equipment and the operation through the analysis and elimination of operational losses (**Table 1**).

Pillar of specific improvements		Evaluation/progress/criteria
Evaluation/progress/criteria	**Background/objective**	**Expected condition**
1. Elaborate complete and detailed flow of the operation, identifying the various auxiliary engines and equipment, their respective priorities, and main risks. Note the current conditions so that you can compare after the improvements are implemented	To increase the knowledge of the whole operation and to standardize the knowledge of the participants of the working group, using the tools of quality	Working group formed, operational flow completed in a clear and didactic way, equipment, priorities, and main risks identified and being known by all participants
2. Identify the generation capacities in MW of each engine/plant—standard and real—and the current losses of the operation, quantifying through the Pareto chart	Identify the distortions between the actual and expected (standard or standard) of each engine/plant. Identify fuel and lubricant/engine/plant consumptions, knowing the performance of each one to be able to act on improvements	Motors, auxiliary equipment, and operations identified with their nominal and actual capacities Criterion to analyze and identify the main losses of the operation, stratify, and classify graphically in A, B, and C (Pareto)
3. Investigate losses in detail according to the priority grades I, II, and III of the chart, presenting alternatives for reducing or eliminating current losses found for later comparison	Allow to identify the fundamental causes of each selected loss, the actual operating conditions of each motor/auxiliary equipment (clearances, paint, leaks, instrumentation, working environment conditions, qualification of operators, necessary and available tools, etc.)	Use of the MASP tools to analyze and solve identified losses. PDCA, Fishbone, 5W2H Methodology being used to investigate and eliminate losses
4. Prepare detailed action plan for the chosen losses, and develop a schedule of activities, following the MASP methodology	Organize the various activities necessary to eliminate identified losses, in order of priority (from highest to lowest) and investment (from lowest to highest)	Plan of action prepared by the working group with actions, responsibility, deadlines, and progress of the activities chosen in the item above, through the MASP tools Put the action plan into practice, and compare the results before and after
5. Standardize operational procedures, ensuring that engines and auxiliary equipment are operated within the required conditions of pressure, temperature, speed, rpm, etc.	After achieving the expected results, standardize the procedures that should be followed by all operators	Interim operational standard completed and being used by the operators in each engine and auxiliary equipment
6. Analyze the existing operational reports, and make the necessary modifications to improve the quality of the annotated information, including maintenance stops by motor or auxiliary equipment, lack of spare parts, labor problems, transportation, etc.	Improving the quality of information to assist in the investigation of losses and their eliminations	Performance of the operation/motor and auxiliary equipment being evaluated by comparing the indicators and objectives defined for each engine/plant. Information of the operational reports being provided with quality and accompanied by the managers, supervisors, and operators. No data distortion

Source: Authors.

Table 1.
Pillar of specific improvements.

3.2 Automatic maintenance pillar (recommended group: managers, supervisors, and operators of each plant)

Objective: To enable the operators to keep their workplaces clean and organized, inspecting their equipment, following operating procedures, lubricating and identifying abnormalities, and labeling and attempting to eliminate hard-to-reach places and sources of dirt (**Table 2**).

Automatic maintenance pillar		Valuation/progress/criteria
Evaluation/progress/criteria	**Background/objective**	**Expected condition**
1. Determine the procedure and how to identify abnormalities through labels. Determine labeling procedures, label types, and colors 2. Train participants to identify abnormalities of motors, auxiliary equipment and work area through labeling 3. Perform initial cleaning on all motors, auxiliary equipment, and operating areas, determining ideal working conditions (no leakage, good flooring, motors, auxiliary equipment and facilities, painted and corrosion-free, with necessary signaling, conditions security, etc.)	Eliminate abnormalities of motors, auxiliary equipment, installations, workplace, and accumulated dirt, eliminate unused materials in the operation, visually identify the abnormal conditions that need to be repaired, and maintain the ideal working conditions that meet Industrial Safety. In this initial cleaning, the conditions of motors, auxiliary equipment, and installations such as loose bolts, lack of fixings and protections, damaged parts and temporary repairs, lack of signaling, etc., identified each with a label, and providing the necessary repairs must be observed. The label must only be removed after approval of the service performed	At initial cleaning, operators and personnel involved must be trained to identify abnormalities in motors, auxiliary equipment, facilities, and workplaces through stickers Areas, engines, auxiliary equipment, and facilities must be clean and maintained in this condition, no longer tolerating any signs of clutter and dirty locations. Use the 5S Locations that are not meeting this requirement should at least be flagged and their future repair be included in a timely, responsible action plan
4. Prepare planning/schedule to carry out the necessary activities of removal of the labels placed in places that presented abnormalities	Monitor the activities performed, and measure the results after the improvements implemented	After label placement, a control should be created indicating the type of problem, the number of labels placed and removed, and the areas involved in the abnormalities, such as maintenance, operation, safety, and environment. Identification, simple, and objective control
5. Establish the basic conditions of engines, auxiliary equipment and facilities, workplaces, floors, walls, lighting, painting, signaling, temperatures, etc.	Ensure operation within the ideal standards required	Ideal conditions for motors, auxiliary equipment, signed installations, and work areas, with industrial safety colors, nameplates, lighting, and cleaning

Source: Authors

Table 2.
Automatic maintenance pillar.

3.3 Planned maintenance pillar (recommended group: PM coordinator, service managers and supervisors, and each plant)

Objective: To create a maintenance management corporate model for all engines and auxiliary equipment of the plants and external clients and to optimize interventions and reduce maintenance costs, ensuring the performance of auxiliary engines and equipment (**Table 3**).

3.4 Pillar of education and training (recommended group: this pillar is corporate and only depends on HR)

Objective: To support the other pillars, analyzing the qualification of participants and the need for training. Responsible for communication, TPM disclosure, event planning, and compliance with the basic program guidelines to facilitate documentation, reduction of dissemination costs, and support material (**Table 4**).

Planned maintenance pillar		Evaluation/progress
Evaluation/progress/criteria	**Background/objective**	**Expected condition**
1. Elaborate and approve methodology to prioritize engines, auxiliary equipment, and facilities in A, B, and C and disclose to all OPM coordinators. Determine form of identification and approve with the steering committee	Standardize how to prioritize engines, auxiliary equipment, and facilities as the company needs, with a focus on business	Complete prioritization worksheet containing pertinent questions from the areas involved in the operation (operation, maintenance, engineering, safety, and environment)
2. Determine how and when the meeting involving operation, engineering, maintenance, safety, and environment will be made to define all the engines, auxiliary equipment, and facilities of each plant in A, B, or C	Identify the company's business priorities to facilitate the deployment of a maintenance management model	Meeting to evaluate and classify in A, B, and C all engines, auxiliary equipment, and facilities of the company, marked or performed with the areas involved
3. After completion, visually list and identify priorities A, B, or C to facilitate supervision	Facilitate service and decision in the most appropriate action to be taken, according to priority	All motors, auxiliary equipment, and facilities, classified in A, B, and C with the visual identification labels, according to the model approved and adopted by the company
4. Identify the current state of each engine, auxiliary equipment, and installation, inspect/review, and make necessary repairs to maintain in perfect operational conditions	Rescue the ideal operating conditions of engines, auxiliary equipment, and facilities, improving availability, reliability, and maintenance	Inspection/revision planning in engines, equipment, and facilities A to redeem desired conditions Action plan defined with activities, materials, deadlines, time provided in each repair activity, and maintenance team
5. Elaborate the most indicated maintenance procedures for each engine, auxiliary equipment, and facilities, as recommended in the master plan Consider those in the technical manuals, MaMa2i, and create those that do not exist and are necessary	Define a maintenance philosophy to be used in equipment A, B, and C according to priority	Equipment A, B, and C classified and with the type and recommended maintenance plan completed

Planned maintenance pillar		Evaluation/progress
Evaluation/progress/criteria	**Background/objective**	**Expected condition**
6. Start the required activities for each engine, auxiliary equipment, and installation A. Follow the MaMa2i plan, and add non-existing services to the system	Organize maintenance, and update the data of each engine, auxiliary equipment, and installation A, creating history, technical inspection standards, and maintenance procedures Follow template created by engineers for reference	Planning and schedule of activities for engines, auxiliary equipment, and facilities A, completed and started

Source: Authors

Table 3.
Planned maintenance pillar.

Pillar of maintenance education and training		Evaluation/progress
Evaluation/progress/criteria	**Background/objective**	**Expected condition**
1. List all employees who have already been trained and those who require basic OPM training to participate in the work groups	Level the knowledge of all participants before starting to develop the activities in the working groups	All employees participating in the OPM program identified to receive the basic training provided by the pillar coordinators
2. Elaborate and make available in the network a basic training to minister to all the employees and in the integration of new ones	Standardize the material and information passed to employees	Teaching material for the basic training, completed, approved by the steering committee, and made available to the pillar coordinators
3. Determine the dates of the training of each pillar and the person in charge of ministering	Organize a schedule of activities to monitor, and audit the development of the TPM	Planning/schedule of training to be performed, indicating employees, dates, and instructors TPM training for integration of new employees, completed to be incorporated by HR
4. After the training, disseminate the number of participants to serve as an evaluation indicator of the pillar in the TPM program	To measure the degree of PMS development and to present the ET pillar indicators	Constant and updated dissemination of the number of employees trained and hours of training performed
5. Make competency map of all participants in the working groups, to identify the qualification, knowledge, and needs	Identify the need for training, planning, and implementation in order to allow the activities of the other OPM pillars to proceed	Worksheet of skills and qualification of the maintainers and operators completed, indicating the basic and specific training required
6. Elaborate internal/external training plan, one-point training, MASP, and lectures to adapt the knowledge need of each work group participant	Level the knowledge of the working group participants so they can take on other activities without any problem, according to the steps of the ME, MA, MP, and SMA pillars	Planning to carry out the training identified in the previous item, including lectures, one-point training, MASP, etc.

Source: Authors

Table 4.
Pillar of maintenance education and training.

Activity name	Responsibility	Check list
Check cleaning of areas	Plant manager	Check leaks, state of conservation, paint, and signage
Check the binder where the "cleaning pattern" of the area is located	Supervisors	Check if the cleaning pattern plug is placed in an easy to read location
Check if the tags are in control	Plant manager	Check in the label control software if there is any movement of placement of new labels
Check action plan to remove labels	Plant manager and supervisors	Check in the label control software if there is an action plan for removing the labels
Check signage of plants	Plant manager and supervisors	Check standardization of signaling
Check GPM frame of the plant	Engineering	Check if the information is up to date
Source: Authors.		

Table 5.
Check list pillar auditoria.

Members of the audit pillar checklist should meet monthly to discuss the MTBF goals and the monthly MTTR and other activities corresponding to the maintenance management program (**Table 5**).

4. Predictive maintenance using computational (fuzzy logic) decision support tool in preload dispatch

An application of fuzzy logic is justified by the ability to anticipate the possibilities of making the pre-dispatch time of the load on the operational tasks of the equipment.

This study deals with the application of fuzzy logic to load dispatch, but with a particularity that is to perform the said pre-dispatch of load taking into account the technical state of the engines, evaluated by different variables related to maintenance. In the first part, the development of the fuzzy rules and of the whole procedure of inference is exposed, and in the second part, all the tests evaluate the maintenance and the technical state of the motors. This tool served as the basis for the resolution of the real problem of pre-dispatch of cargo to satisfy the rationalized methods of just in time of the thermal plant on the operational conditions of the equipment (**Figure 2**).

A system based on fuzzy logic, as shown in **Figure 2**, can have its action schematized by the following constituent elements: fuzzifier; rules, or knowledge base; inference, or logical decision-making, and Defuzzifier [13].

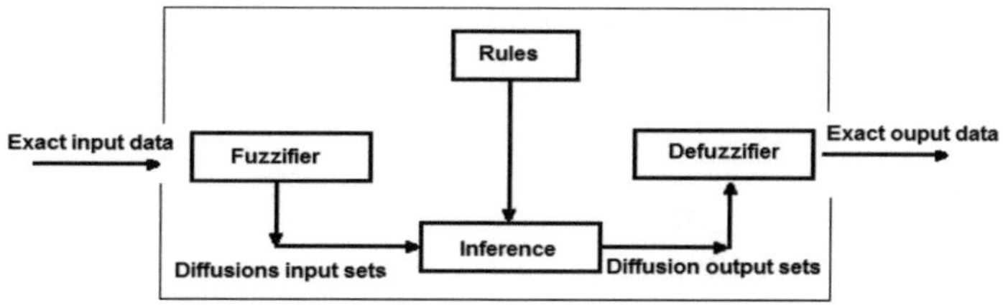

Figure 2.
System based on fuzzy logic. Source: Authors.

In the first part the development of the fuzzy rules and of the whole procedure of inference is exposed and in the second part all the tests to evaluate the maintenance and the technical state of the motors. This tool served as the basis for the resolution of the real problem of pre-shipment of cargo to satisfy the rationalized methods of just in time of the thermal plant on the operational conditions of the equipment [14, 15].

The fuzzy system models the style of reasoning, imitating the capability to make decisions in an environment of uncertainty and imprecision. In this way, fuzzy logic is an intelligent technology, which provides a mechanism to manipulate imprecise information—concepts of small, high, good, very hot, cold—and that is able to infer an estimated answer to a question based on an inexact, incomplete knowledge, or not fully reliable information.

The development of a computational tool supports the load dispatch according to the location of motors and generators for thermal energy, analyzing the main thermoelectric generation variables for the entire predictive maintenance process.

All variables are inserted considering the intervals determined in the rules of inference as shown below.

The computational interface was useful for the search of some preselected characteristics to enable its implementation. **Tables 6–12** show such characteristics and respective purposes.

In this context, the following groups of information and data are abstracted: the input values, called crisp, the linguistic variables, and the fuzzy variables. The fuzzy logic is justified in the solution of this case study in function of the input variables with better representation in fuzzy sets. The variables due to the dimension of the universe of study were divided in 04 (three) and 03 (two) inputs and 01 (one) output, all independent of each other.

4.1 The input variable "vibration analysis"

For the determination of each variable, it was convenient to divide them into strips to approximate the actual situation to be checked. The calculation of these ranges on a scale according to **Tables 6–12** is shown below.

As the first level of variation, "vibration level" in **Tables 6** and 7, let us consider better variable levels that were subdivided into four variables, normal, permissive, alert, and critical, each corresponding to the classification of vibration, velocity, and displacement levels measured in the equipment.

4.2 The input variable lubricating oil

The "level of analysis of the lubricating oil," **Tables 8** and **9**, can be presented, for example, with the water content in the oil and solid and non-lubricated particle

Class	1-[N] Normal	2-[P] Permissible	3-[A] Alert	4-[C] Critical mm/s
(Class I)	(0.18–0.71)	(0.71–1.80)	(1.80–4.50)	(Above 4.50)
(Class II)	(0.18–1.10)	(1.10–2.80)	(2.80–7.10)	(Above 7.10)
(Class III)	(0.18–1.80)	(1.80–4.50)	(4.50–11.2)	(Above 11.2)
(Class IV)	(0.18–2.80)	(2.80–7.10)	(7.10–18.0)	(Above 18.0)
	A	B	C	D

Source: Authors

Table 6.
Manufacturer vibration levels.

Zone	Qualification	Operation of machines
Zone A	[N] Normal 0.18–2.80 mm/s	Commissioned machines should generally operate in this area
Zone B	[P] Permissible 2.80–7.10 mm/s	It is acceptable for unrestricted operation for long periods
Zone C	[A] Alert 7.10–18.0 mm/s	Unsatisfactory for continuous operations for long periods
Zone D	[C] Critical above 18.0 mm/s	It is sufficient to cause damage to the machine at any time

Source: Authors.

Table 7.
Vibration severity rating relevance function.

Class	1-[N] Normal	2-[A] Alert	3-[C] Critical
(Water% volume)	(% ≤ 0.2)	(0.3)	(Above 03)
(Micron iron content)	(% ≤ 49)	(50)	(Above 51)
(Micron copper content)	(% ≤ 1)	(20)	(Above 21)
	A	B	C

Source: Authors.

Table 8.
Lubricating oil.

Zone	Qualification	Operation of machines
A	[N] Normal	Commissioned machines should generally operate in this area
Water% volume	% ≤ 0.2	
Micron iron content	% ≤ 49	
Micron copper content	% ≤ 19	
B	[A] Alert	Unsatisfactory for continuous operations for long periods
Water% volume	0.3	
Micron iron content	50	
Micron copper content	20	
C	[C] Critical	It is sufficient to cause damage to the machine at any time
Water% volume	Above 0.3	
Micron iron content	Above 51	
Micron copper content	Above 21	

Source: Authors.

Table 9.
Function of pertinence of the severity according to the oil.

content (iron and copper), the energy sources of the dispatch of load for generation of energy. The levels of analysis of the command type were subdivided into 03 (three) variables, correspondence and information quality [9].

(Zone)		(Thermography)	
(A)		([N] Normal less or equal 94.0 F)	
B		[P] Permissible	
(B/C)	(94.0 F)		(164.2 F)
C		[A] Alert	
(C/D)	(164.2 F)		(199.3 F)
(D)		[C] Critical above 199.3 F	
Source: Authors			

Table 10.
Thermography to determine hot spots.

Zone	Qualification	Operation of machines
Zone A	[N] Normal (T ≤ 34.5°F)	Commissioned machines should generally operate in this area
Zone B	[P] Permissible (34.5°F < T ≤ 73.5°F)	It is acceptable for unrestricted operation for long periods
Zone C	[A] Alert (73.5°F < T ≤ 93°F)	Unsatisfactory for continuous operations for long periods
Zone D	[C] Critical (T > 93°F)	It is sufficient to cause damage to the machine at any time
Source: Authors		

Table 11.
Function of pertinence of the classification of thermography.

Motor technical status (ETM) for operating conditions		Operation of machines
Normal	76–100%	Commissioned machines should generally operate in this area
Permissible	51–75%	It is acceptable for unrestricted operation for long periods
Alert	26–50%	Unsatisfactory for continuous operations for long periods
Critical	0–25%	It is sufficient to cause damage to the machine at any time
Source: Authors.		

Table 12.
Variable "engine technical status".

4.3 The input variable "thermography analysis level"

Thermography analysis is an input variable, **Tables 10** and **11**, that can be used as a tool for load dispatching. The levels of thermographic analysis were just been subdivided in four (4) variables, each one corresponding to the dynamic memory. The use of images in thermal plants is very important for this reason. The infrared radiation is a base of studies on the thermal images, which has a function of capturing this radiation, interpreting and generating a quantitative image of the temperature of the studied body [16].

4.4 Output variable "technical condition of the motor"

The "estimated technical state of the engine (ETM)" is the output variable of the system, in relation to vibration (oil, water, iron, and copper). **Table 12** describes the operating state of the generating units. The variable under study, as well as the variable "level," was transferred to a percentage scale of 100% where "GREET" corresponds to the range of maximum values and the variable "BAD" corresponds to the range of minimum values up to zero. This value gives a greater range of possibilities, making the case study more precise.

5. Fuzzy simulation

The fuzzy inference with the input and output variables was performed employing the MATLAB version 8.0 tool and using a Mamdani model. This model adopts semantic rules for the processing of inferences and is commonly referred to as maximum-minimum inference. Such an inference model applies well to this type of problem since it uses union and intersection operations between sets. All variables are entered considering the intervals determined in the rules of inference. **Figure 3** shows the variables "vibration," "water," thermography, "iron," and "copper" according to **Figure 3**.

All variables are entered considering the intervals determined in the rules of inference. **Figures 4–7** show the variables "vibration, water, thermography, iron, and copper" according to the figures below.

The first input variable is a thermography (**Figure 4**). According to **Tables 6** and **7**, we have

The second input variable is water (**Figure 5**) produced by the generating units. According to **Tables 8** and **9**, we have

The third input variable is the thermography (**Figure 6**) produced by the generating units. According to **Tables 10** and **11**, we have

The fourth input variable is iron (**Figure 7**) produced by the generating units. According to **Tables 8** and **9**, we have

The fifth input variable is copper (**Figure 8**) produced by the generating units. According to **Tables 8** and **9**, we have

The motor technical state is a product of the relationship between the input variable and output variable, which compose the pertinence functions expressed in the curves of **Figure 9**.

After editing the pertinence functions of all variables, the implemented rules are arranged in **Figure 10**, as shown in **Figure 8** for the visualization of the linguistic variables, thus forming antecedents and subsequent ones based on the Fuzzy inference rules .

To better understand the screen expressed, **Figure 11** shows all the possibilities that the simulation can produce. The movement of the red lines determines the other rule to be evaluated.

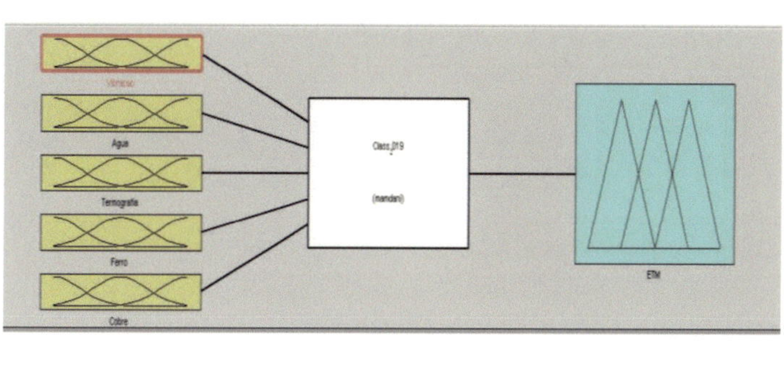

Figure 3.
Mamdani's model. Source: Authors.

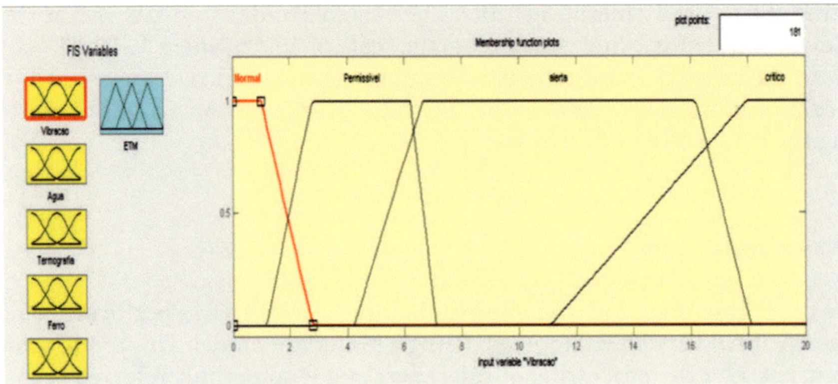

Figure 4.
"Vibration level." Source: Authors.

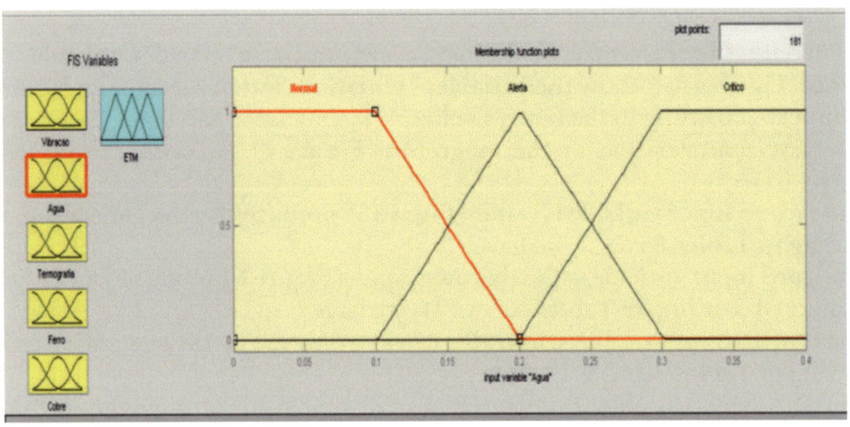

Figure 5.
"Water." Source: Authors.

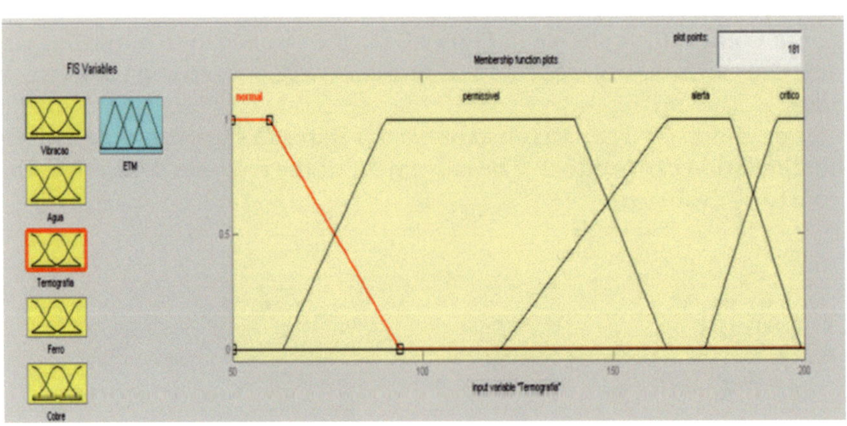

Figure 6.
"Thermography." Source: Authors.

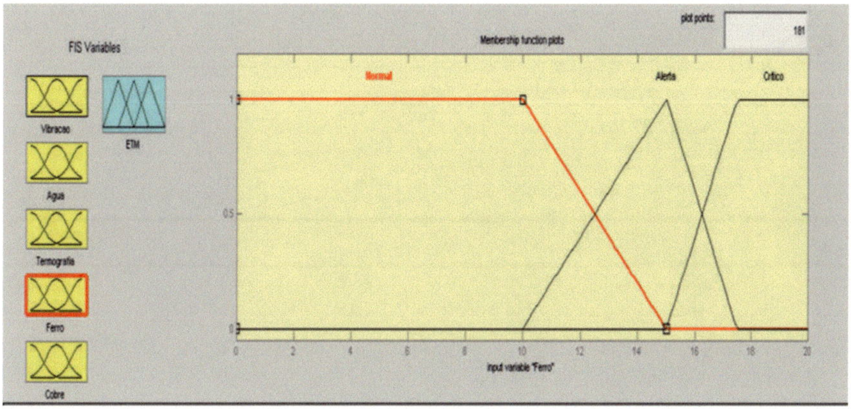

Figure 7.
"Copper." Source: Authors.

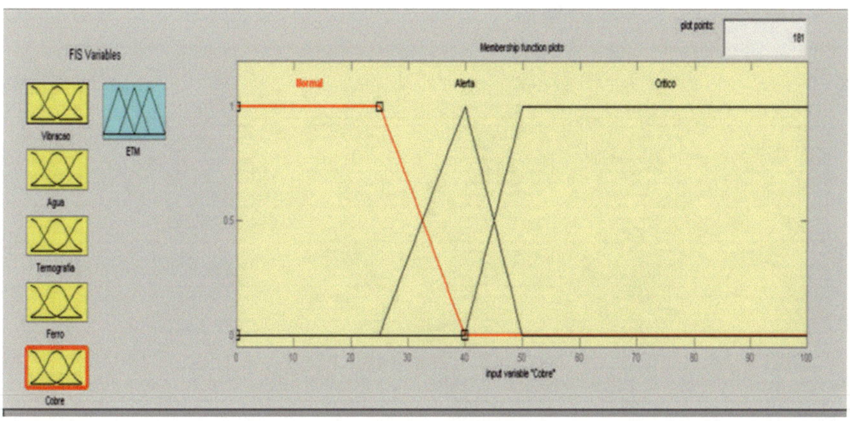

Figure 8.
"Copper." Source: Authors.

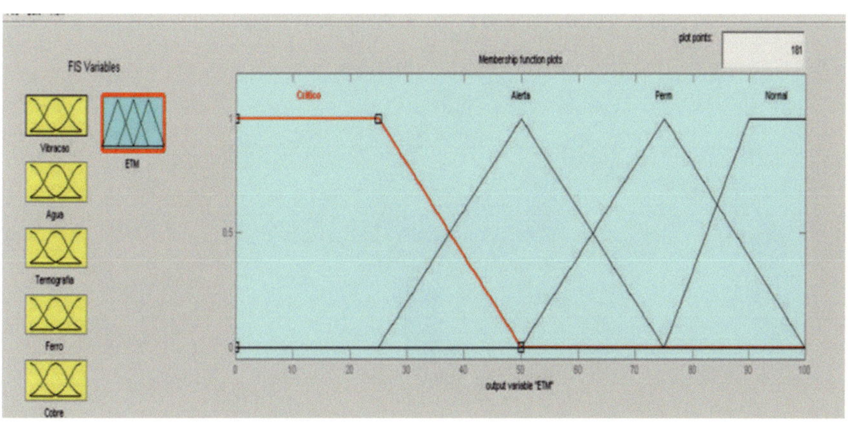

Figure 9.
Output variable: technical status. Source: Authors.

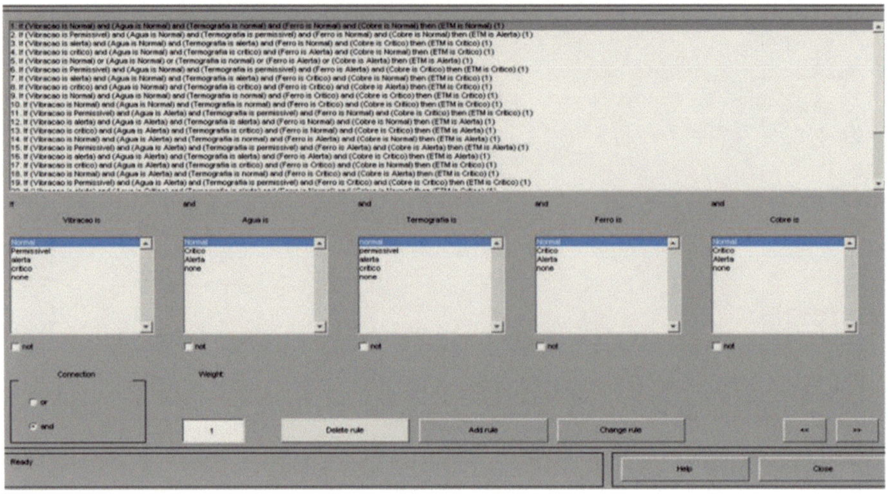

Figure 10.
Implemented inference rules. Source: Authors.

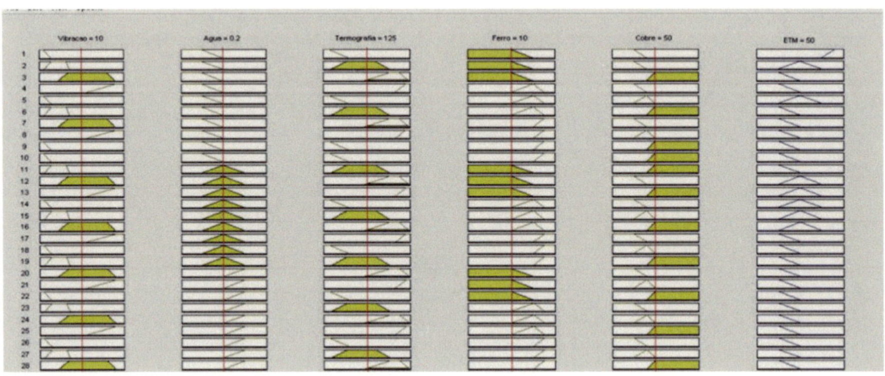

Figure 11.
The input and output variables. Source: Authors.

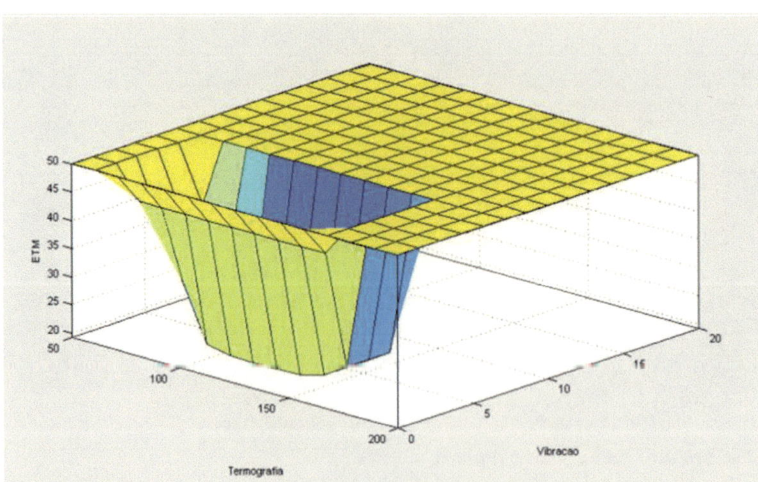

Figure 12.
(Thermography × vibration). Source: Authors.

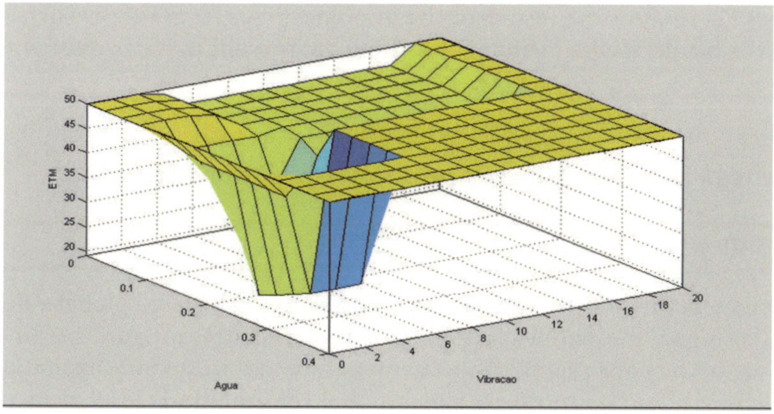

Figure 13.
(Water vs. vibration). Source: Authors.

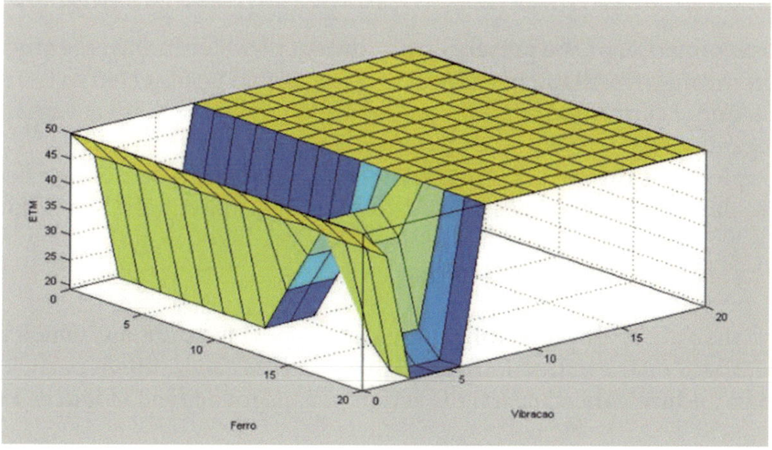

Figure 14.
(Iron × vibration). Source: Authors.

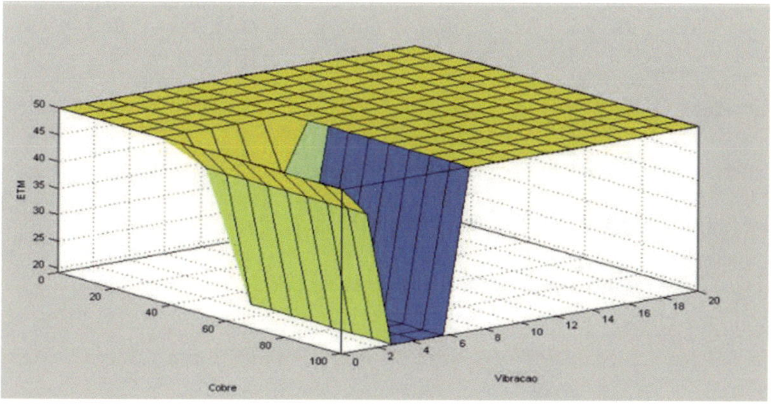

Figure 15.
(Copper vs. vibration). Source: Authors.

Figures 12–15 show the results of the inference rules from the 3D surface of the graph. In the blank, all the forms of execution are present that can exist within the simulation.

6. Case study (fuzzy logic with predictive maintenance).

6.1 Vibration analysis

Equipment status control is performed based on a calculated global value for the vibration signal measured at critical points on the machine surface. Since this value is due to a response signal from the structure to the dynamic excitation of the equipment operation, it represents a measure of the amplitude level of its vibration signal. In the case of the application for predictive maintenance, the international technical standards, among them the ISO, define two criteria for adoption of a global value (**Figure 16**).

6.2 Analysis of water content in lubricating oil

The determination of the presence and content of water in the case study was carried out through the distillation by drag. The sample is subjected to heating for distillation under controlled conditions, thus verifying the water content in the lubricating oil.

The graph shows the results of the analysis of water content, done periodically as predictive, showing normal levels, since the tolerable content is 0.3% (**Figure 17**).

6.2.1 Analysis of metal content in lubricating oil

The graph made by direct reading (iron and copper) ferrography, which was carried out based on the extraction of the magnetizable contaminant particles, is contained in the lubricant, through the action of a magnetic field (**Figures 18** and **19**).

6.3 Thermography

In addition to the use of the supervision system provided by the 9 FLUKE software, a thermovision is used, as shown in the figure, for measurement in low

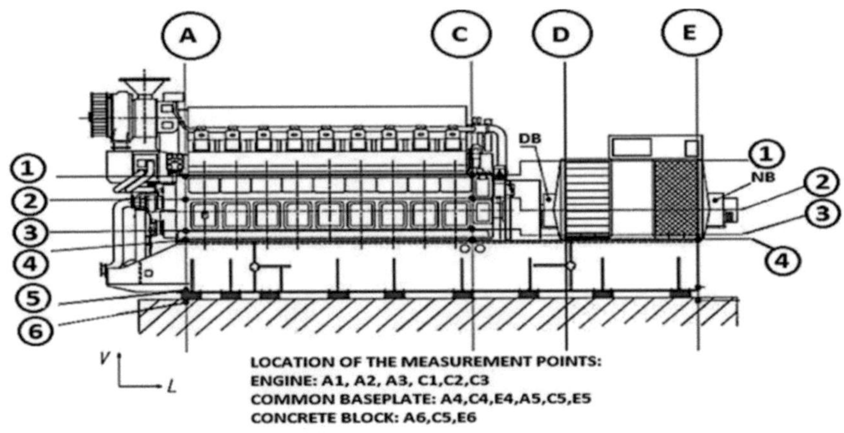

Figure 16.
Measurement points in the vibration analysis. Source: Authors.

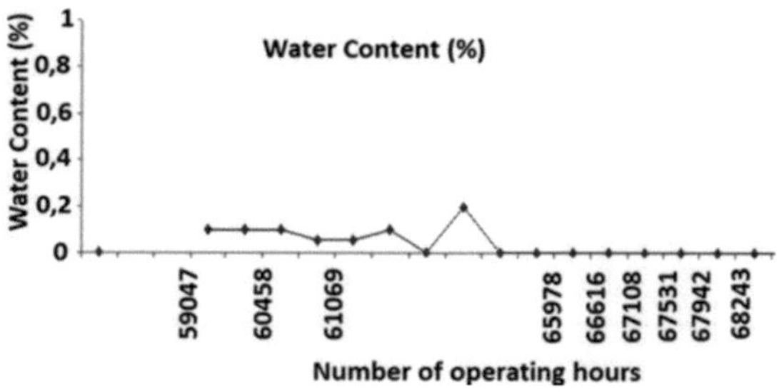

Figure 17.
Water content and lubricating oil. Source: Authors.

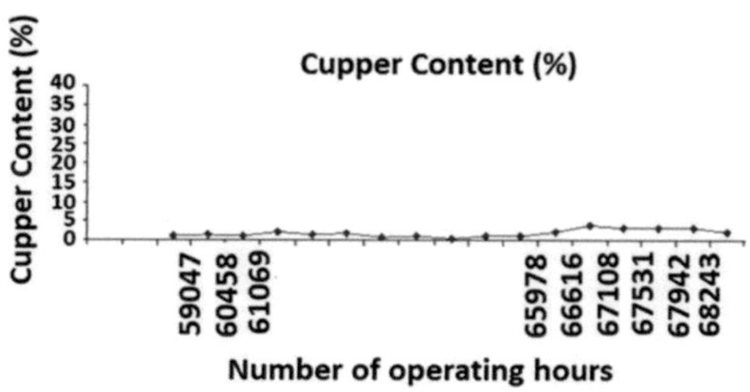

Figure 18.
Copper content in the lubricating oil. Source: Authors.

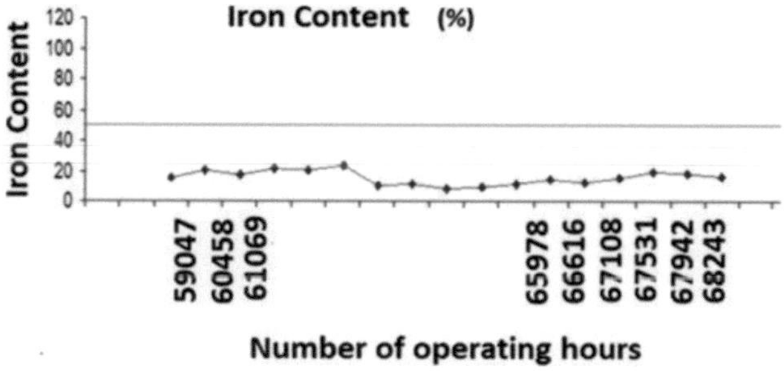

Figure 19.
Iron content in the lubricating oil. Source: Authors.

or high voltage electrical systems, temperature variations caused by excess electric current in the furnace motor/generator 01 be with the hot spot and it will not be able to pre-dispatch cargo, **Figure 20**.

6.3.1. Fuzzy logic goes into the operating conditions of the equipment

In **Figure 21**, we can identify the anomalies likely to occur in the electric motor of generator 1 for all effects of temperature caused by excessive electric current. The heating screen of **Figure 22** indicates that the engine/generator 1 cannot be related for preloading under operating plants. Other motors and generators are in the normal comfort area (A) and can be classified for normal operation of the diffuse rule such as the motors and tuners 2, 3, 4, 5, 6, 7, 8, 9, and 10. 5 and 25 mark as normal operations without interruption and execution of restriction only for the motor/generator 1. Activate the excluded points in your electronic connection.

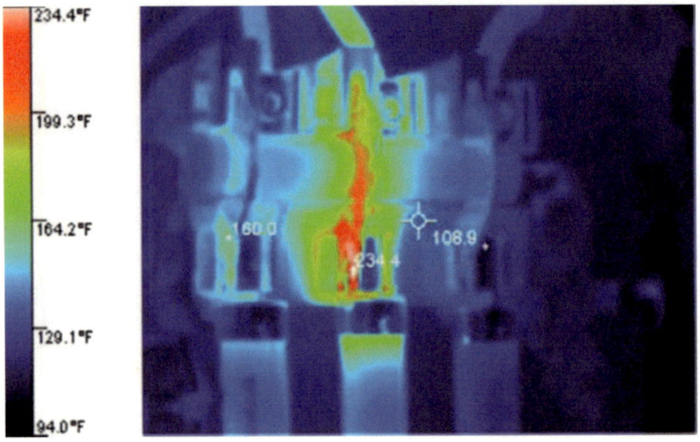

Figure 20.
Copper content in lubricating oil. Source: Authors.

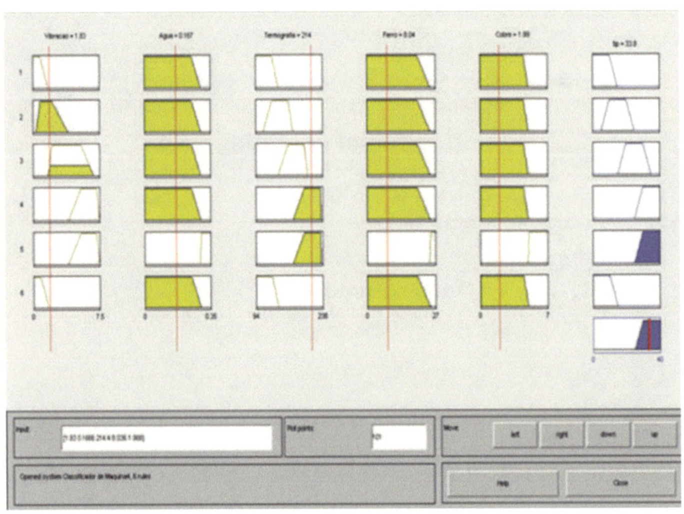

Figure 21.
Copper content in the lubricating oil. Source: Authors.

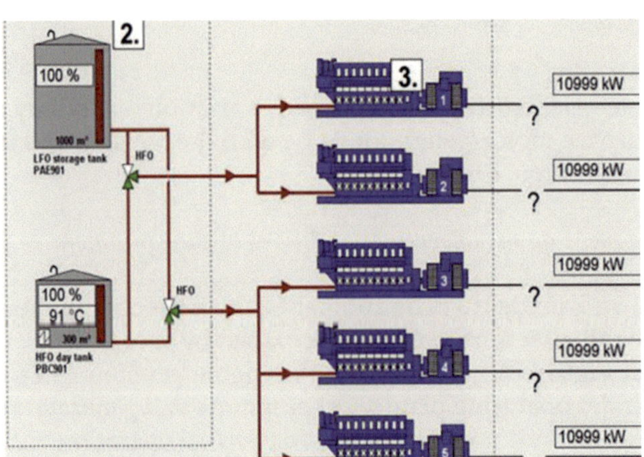

Figure 22.
Copper content in the lubricating oil. Source: Authors.

In **Figure 21** the parameters for the location of the equipment, according to the engine/generator 1, are not allowed to operate because they are not in good operating condition. The other engines and generators are located in area A (N) normal 2, 3, 4, 5, 6, 7, 8, 9, and 10 and are able to position themselves according to the needs of the organization. **Figure 22** informs which engines are conditional ideal for pre-shipment of cargo under operating conditions.

7. Results achieved

The objective of this work was to analyze the maintenance management system and its optimization through nebulous logic for the development of an intelligent system of support and decision-making for an ideal load dispatch demand.

The interface of the developed computational tool achieved the simplicity desired by the users themselves, as well as the ease of learning in their operation. According to the facts presented in this paper, it was possible to show that, currently, a predictive maintenance program and a total maintenance program are indispensable for large companies. This is to provide reliability to processes and equipment, detecting problems still in the initial phase. Programs of this type provide good maintenance planning for the maintenance industry. Thus, the company grows with regard to meeting deadlines, resulting in an increase in customer satisfaction.

In the present study, the gains from the two plans mentioned above could be assessed based on the information from the case study; we verified the reduction of corrective maintenance, and we verified the results with the increase of the MTBF and the decrease of the MTTR. The observed case can be implemented in any power generation machine that uses the fuel oil and, consequently, the use of oil stock, independent of the tank capacity and storage tank scales, which have only the standards of this system. The case study presented here can be implemented in any thermoelectric plant, independent of the loads to be dispatched, since the variables of this system are common to all.

Acknowledgements

The authors gratefully acknowledge the support of this research by FAPEAM, UFPA, ELETROBRAS, and ITEGAM.

Author details

Milton Fonseca Junior[1*], Antônio Estanislau Sanches[5], Jandecy Cabral Leite[2], Jorge de Almeida Brito Júnior[2], Manoel Henrique Reis Nascimento[2], Carlos Alberto Oliveira de Freitas[2], Tirso Lorenzo Reyes Carvajal[2], David Barbosa de Alencar[2], Jorge Laureano Moya Rodriguez[3], Ubiratan Holanda Berzerra[4] and Nadime Mustafa Moraes[5]

1 Engineering Department, Generation of Mauá, Eletrobrás Amazonas GT, Manaus, Amazonas, Brazil

2 Research Department, Institute of Technology and Education Galileo of the Amazon (ITEGAM), Manaus, Amazonas, Brazil

3 Department of Post Graduate in Industrial Engineering, Federal University of Bahia (UFBA), Brazil

4 Department of Post Graduate in Electric Engineering, Federal University of Pará (UFPA), Brazil

5 University of the State of Amazon (UEA), Manaus, Amazonas, Brazil

*Address all correspondence to: milton.fonseca.jr@gmail.com

References

[1] Silva SG. Sistema interligado nacional: análise das penalizações impostas às transmissoras com foco na aplicação da parcela variável. Dissertação (Mestrado em Ciências)-Universidade Federal de Uberlândia, Uberlândia. 2016. p. 122

[2] Ma H et al. Multi-objective biogeography-based optimization for dynamic economic emission load dispatch considering plug-in electric vehicles charging. Energy. 2017;**135**: 101-111. ISSN 0360-5442

[3] Pless J, Fell H. Bribes, bureaucracies, and blackouts: Towards understanding how corruption at the firm level impacts electricity reliability. Resource and Energy Economics. 2017;**47**:36-55. ISSN 0928-7655

[4] Kumar N. A non convex cost function based optimal load dispatch using tlbo algorithm. Journal of Engineering Science & Technology Review. 2017;**10**(1). ISSN 1791-2377

[5] Alam N et al. A 0/1 mixed integer linear programming approach to establish an effective preventive maintenance policy for power plant. International Journal of Industrial and Systems Engineering. 2017;**25**(4):478-498

[6] Alaswad S, Xiang Y. A review on condition-based maintenance optimization models for stochastically deteriorating system. Reliability Engineering & System Safety. 2017;**157**:54-63

[7] Arya R. Ranking of feeder sections of distribution systems for maintenance prioritization accounting distributed generations and loads using diagnostic importance factor (DIF). International Journal of Electrical Power & Energy Systems. 2016;**74**:70-77

[8] Azadeh A, Asadzadeh SM, Salehi N, Firoozi M. Condition-based maintenance effectiveness for series–parallel power generation system—A combined Markovian simulation model. Reliability Engineering & System Safety. 2015;**142**:357-368

[9] Baidya R, Ghosh SK. Model for a predictive maintenance system effectiveness using the analytical hierarchy process as analytical tool. IFAC-PapersOnLine. 2015;**48**(3):1463-1468

[10] Ripper is a 2001 Canadian-British horror film, directed by John Eyres. The film stars A. J. Cook and Bruce Payne. It was written and produced by John A. Curtis

[11] Fonseca, Miltonx, et al. Pre-dispatch of load in thermoelectric power plants considering maintenance management using fuzzy logic. IEEE Access. 2018;**6**:41379-41390

[12] Junior MF et al. Maintenance tools applied to electric generators to improve energy efficiency and power quality of thermoelectric power plants. Energies. 2017;**10**(8):1091

[13] Larguech S et al. Fuzzy sliding mode control for turbocharged diesel engine. Journal of Dynamic Systems, Measurement, and Control. 2016;**138**(1):011009. ISSN 0022-0434

[14] Gonçalves BP et al. Avaliação de impactos harmônicos na rede elétrica através dos indicadores THD e fator de potência utilizando lógica fuzzy. Revista Brasileira de Energia. 2013;**19**(1):9-27

[15] Nogueira EL, Nascimento MHR. Inventory control applying sales demand prevision based on fuzzy inference system. ITEGAM-Journal of Engineering and Technology for Industrial Applications (ITEGAM-JETIA). 2017;**3**:31-36. DOI: 10.5935/2447-0228.20170060

[16] López-Pérez D, Antonino-Daviu J. Application of infrared thermography to failure detection in industrial induction motors: Case stories. IEEE Transactions on Industry Applications. 2017;**53**(3):1901-1908

Chapter 5

Maintenance Management of Aging Oil and Gas Facilities

Riaz Khan, Ammeran B. Mad, Khairil Osman and Mohd Asyraf Abd Aziz

Abstract

Oil and gas operators are now being driven to now operate beyond their originally conceived design life and field life. Asset life extension (ALE) beyond these thresholds presents unique safety and business risk challenges for the oil and gas industry. With aging equipment and facilities, operators face increasing challenges in maintaining equipment reliability and integrity as well as operational safety. Aging factors do not only involve hardware but also human and organizational factors. Factors include corrosion, erosion, fatigue, equipment obsolescence, normalization of deviance (accepting degraded conditions as being normal), changes in codes and standards and lack of data to forecast future risks. The challenge is magnified if there is a large fleet or large amount of aging assets that needs to be managed. In this chapter, a responsible approach to ALE, where assets can continue to be operated safely and resources are adequately managed, is provided herein.

Keywords: maintenance management, asset life extension, integrity management, safety critical element, oil and gas producers

1. Introduction

Oil and gas facilities range from both upstream and downstream assets to include offshore structures, onshore tank farm facilities. Offshore structures may include the typical fixed offshore structures, monopods, guyed wire caissons to the more complex deep water assets including Floating Production and Storage Offloading (FPSO), Mobile Offshore Production Unit (MOPU), Tension Leg Platform (TLP) and semi-submersible structures (**Figures 1** and **2**).

Extending operation facilities beyond design life presents safety risks, business risks and operational challenges to the oil and gas industry. These risks affect significant business decisions and need to be quantified and managed as we strive for continuous operations of aging assets. Aging assets and equipment present increased challenges in maintaining equipment integrity and hence, will need to be managed accordingly. These could be because of a cumulative degradation and risks over time, which includes:

- Degraded materials of construction due to corrosion related mechanisms;
- Erosion, wear, fatigue or cracking mechanisms;

- 'Slow burn' degradation mechanisms;
- Obsolescence of equipment leading to potential lack of spares, high cost of spares, etc.;
- Normalization of deviance associated with human factors (i.e. accepting degraded conditions as being the new normal);
- Lack of data trending to forecast future risks to safety and business continuity;
- Failure to record the accurate status of safety critical elements (SCE) over time;
- Changes to engineering codes and standards;
- Loss of technical competence (qualifications + training + experience) in the industry;
- Introduction of foreign materials into the production systems (e.g. Chemicals for Enhanced Oil Recovery (EOR), downhole sand consolidation, chemical tracers, off spec water injection, etc.).

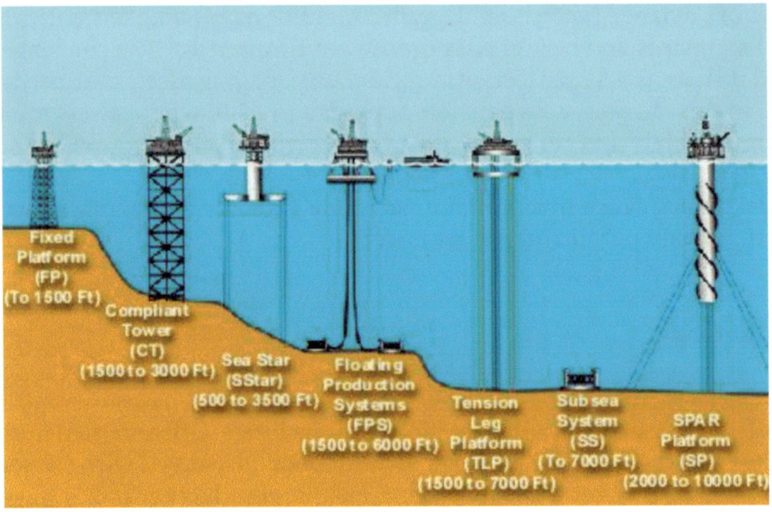

Figure 1.
Typical offshore structures [1].

Figure 2.
Typical onshore tank farm facility [2].

Assets are required to predict and understand the effects of deterioration, or changing conditions associated with life extension and be prepared to intervene to ensure that this demand can be met without adverse effect on asset integrity and safety. Asset life extension (ALE) for a given design life expiry, refers to a condition whereby an asset is approaching its intended design life. The main aging factors that need to be considered when developing an ALE program are material degradation (**Figure 3**), obsolescence and organizational issues. This is provided within **Figure 4**.

The status of the known degradation mechanisms applicable for safety barriers should be evaluated and documented. The basis for acceptance of deviations and management of change (MoC) is reviewed in as a justification for the new mode and timeframe for continuous operations. The engineering evaluations of all changes and eventually mitigation measures against all operating risks must be documented. OGPs must review, evaluate assess all damage mechanisms or defects that may impact the facilities or individual operating systems for the life extension period. This is generally applicable to damage or defects where a temporary MoC has been accepted due to a limited period of use and this period has since been changed as a result of ALE considerations. The OGP is then required to re-assess the basis for acceptance to verify that this is still valid for the new period. Components or systems with a high consequence of failure, which are not available for inspection must be identified, evaluated, analyzed, and qualified for life extension. It is required that OGPs evaluate the consequence in case of failure, monitors

Figure 3.
Degradation of offshore structural component [3].

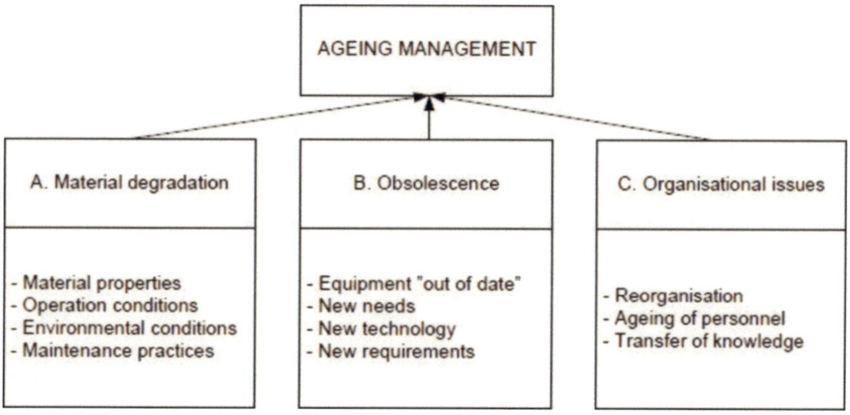

Figure 4.
Aging management [4].

indications of failure and have plans for compensating actions if indications of failure are found. Latest knowledge related to degradation and life extension shall be applied.

A case study is provided within Section 11, to demonstrate a simple application of the ALE framework and possible outcomes.

2. Operational context

As the Asset ages, there is increasing challenge to maintaining equipment and installation integrity, compliance with Regulatory requirements and improve economic hydrocarbon recovery from depleted fields. As such, life extension analysis and evaluations must be based on the planned use of the facilities during the life extension period. Changes to the operational conditions that can have an impact on the efficiency of resource exploitation, the risk profile as well as the performance of the barriers due to aging, must be considered. The potential changes to the operational conditions that influence the degradation of barriers must be identified and used as basis for life extension evaluations.

Based on Norwegian Oil and Gas Recommended Assessment and Documentation for Service Life Extension of Facilities, Rev1, 2012 [5] and operational data and requirements, the following should, among others, be considered:

- reservoir depletion causing subsidence of the facility
- shallow gas detection and mitigation
- changes in climatic conditions resulting in changes to environmental loadings and operating conditions
- Increased changes in fluid compositions that can adversely affect the corrosion rates in certain systems
- Changes to the original design assumptions as provided in QRA etc.
- Well and drilling factors
- Plans for increased gas flow
- Need for new process or utility equipment due to changed flows, chemistry, pressure, injection or chemicals
- Changes to the SCEs on the facility
- New methodologies to simulate damage and degradation.
- Changes to equipment usage.

3. An asset life extension program

The basis for the design and design life of facilities with its associated platform, wells, subsea systems and pipelines may be different. When facilities are planned to be used beyond design life, OGPs should define the life extension period for which the different parts of the facilities are planned to be used. An ALE framework outlining the main tasks as a six (6) step process is proposed and provided below on **Figure 5**.

3.1 Data and information

The collection of data and information is often the most challenging aspects of commencing an ALE study. It is recommended that records be securely placed within an electronic database generally used to manage asset integrity and reliability solutions. The availability and accuracy of information should be evaluated for each facility considered. The information should constitute design basis and specifications, design and as built drawings, design/(re-) analysis reports, inspection reports, maintenance and repair records and specifications. Once these records have

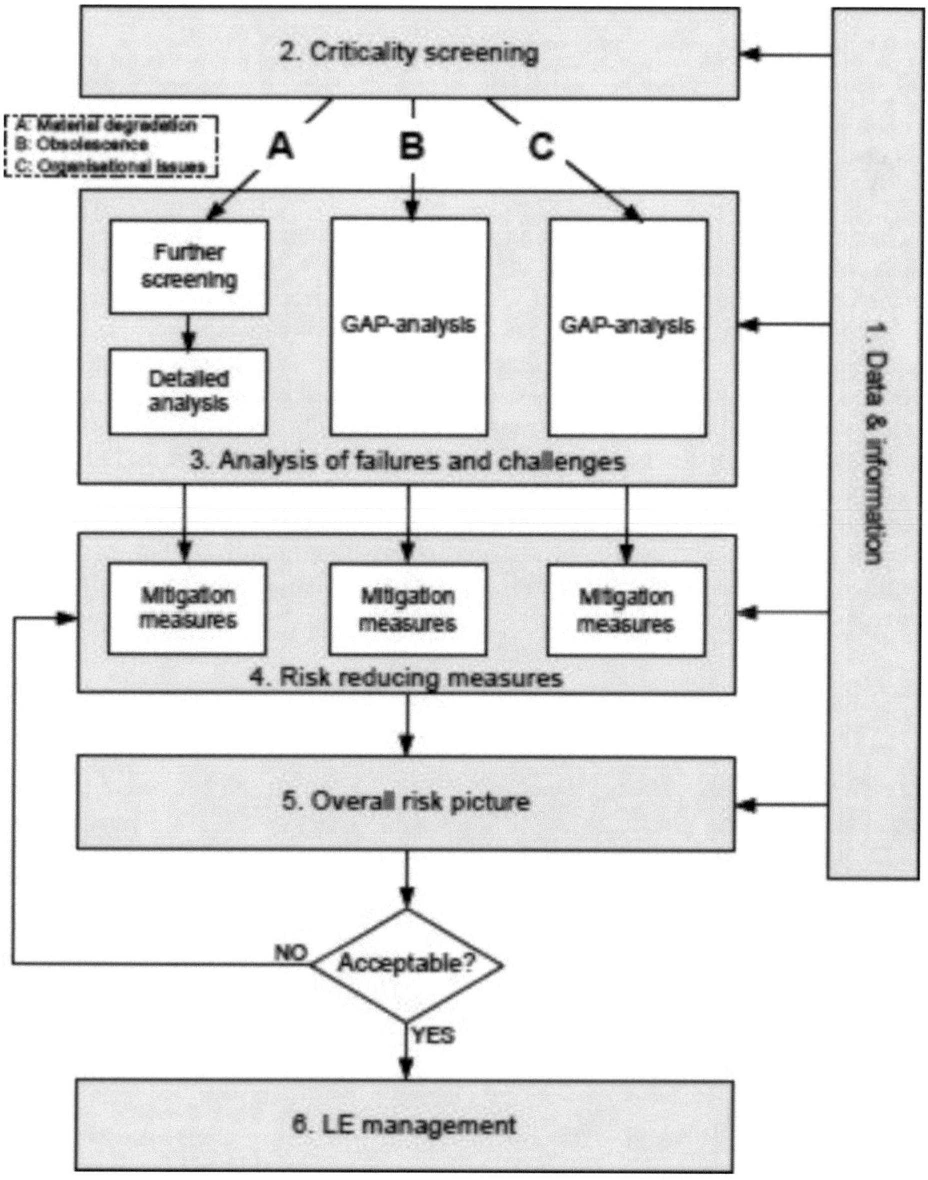

Figure 5.
Asset life extension process [6].

been complied, data quality measures should ensure the appropriate data for screening. In some cases, the data analytics and trending measures give a better representation of the data set and how this can be used effectively in an ALE program.

3.2 Gap assessment

Gap assessment is the second stage of an ALE process. Identifying gaps can be broken down to several steps, which includes:

- Identify hazards and critical barriers.
- Check integrity and functionality of barriers.
- Assess current performance of barrier against intent.
- Review historic performance of barriers.
- Review current state of maintenance and gaps.

The gap assessment shall focus on the barrier functions and the factors that influence the barrier elements. This includes technical, organizational and operational elements. The gap assessment and recommendations are performed based on the major inspection findings, root cause failure analysis reports, modification implemented on the equipment, bad actor list, history of incidents, maintenance report, overhaul findings, reliability data, operating and maintenance philosophy and any condition monitoring recommendation. Any life extension recommendation must take the future technical condition, operating parameters and mode of operation into consideration. The assessment should also include review of the forecasted production profile, exploiting synergy with other related equipment such that key assets and system infrastructure can be rationalized, optimized or expanded. A process for the identification and correction of gaps is provided within **Figure 6**.

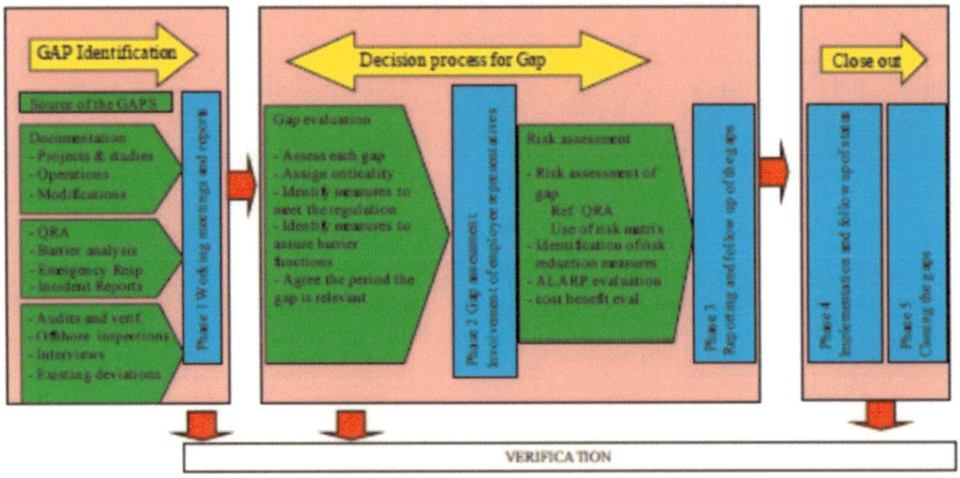

Figure 6.
Process for identification and correction of gaps [5].

The recommendations from the gap assessment are to cover all the remedial actions necessary to prevent the risk associated with spare strategy, obsolescence related to the equipment and spare parts, remnant life analysis and prediction of future failures modes and degradation mechanism especially related to aging during the extension period. The benefits of applying new technology in addressing the gaps shall be evaluated. This could help mitigate or close gaps with less modifications or compensating measures. The Health and Safety Executive, UK (2013) KP4 Report [7] outlined the following safety management systems as being the barriers on the facilities that are not to be breached. They include:

- Structural integrity;
- Process integrity;
- Fire and explosion;
- Mechanical integrity;
- Electrical, control and instrumentation;
- Marine integrity;
- Pipelines;
- Corrosion;
- Human factors.

In addition to the above mentioned the following systems may be considered

- Cranes and lifting equipment
- Telecommunication facilities
- Subsea systems
- Life-saving equipment

Oil and gas producers (OGPs) are to perform analyses and evaluations to demonstrate and understand of how the time and aging processes will affect HSE, the facilities barriers including technical operational and organizational aspects and resource exploitation. They shall also identify measures required to mitigate the impact of the time and aging processes (**Figure** 7).

The Norwegian Oil and Gas Recommended Assessment and Documentation for Service Life Extension of Facilities, (2012) [5] provides good guidance on the processes, resources and methodologies used in the ALE approach to find the "as is" condition and re-qualification for life extension and how to implement and document. Safety critical elements (SCEs) such as wells, subsea jacket structures, pipelines, risers, mechanical equipment etc. are to be qualified for the continuous operations and asset life extension. Quantitative and qualitative assessments are generally employed for equipment where known degradation mechanisms are prevalent and where quantitative models exist to calculate degradation, remaining margins and prediction of remaining service life. Quantitative analysis including

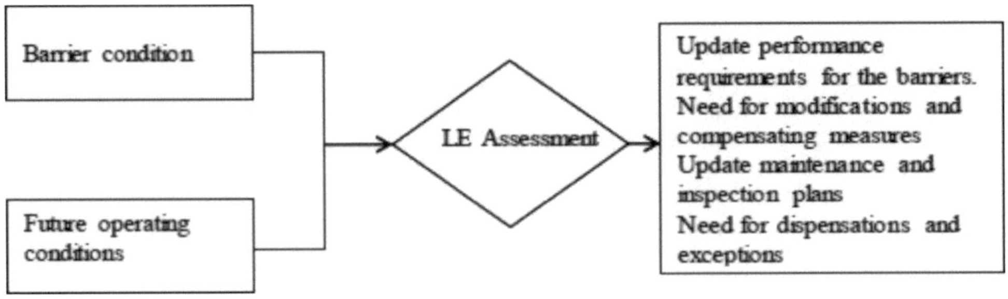

Figure 7.
Recommended life extension assessments of barriers [5].

probability of failure (PoF) is generally employed for structures, pipelines, position mooring, and flexible or steel catenary risers etc. and requires string technical expertise and often specialist software packages. Qualitative assessments is also possible but must be supported by effective data management and operating historical data to make good engineering assessments.

3.3 Risk factors and assessments

Risk assessments must be performed to verify that the facilities risk level is within acceptable limits in the period of life extension and As Low as Reasonably Practicable (ALARP). The principle of ALARP is in widespread use in the oil and gas industry. The following risk evaluations shall be performed based on the context defined for life extension:

- Accumulation of Operational Risk Assessments (ORA), as some of which may be decoupled because they have been considered in isolation and not in combination, potentially resulting in unknown increased risks
- Risk assessment of major accident risk, Quantitative/Qualitative Risk Analysis (QRA)
- Emergency preparedness and response
- External environment
- Occupational safety, health and working environment.

Ensuring risks have been reduced to ALARP means balancing the risks against the costs to further reduce it. The decision is weighted in favor of health and safety because the presumption is that OGPs should implement the risk reduction measure. It is expected that the latest available technology and knowledge related to analysis of major accidents is applied. The conservatism level and any assumptions made in risk assessments are to be assessed and evaluated for all continuous operations. The vulnerability, actual and expected effectiveness of the barrier function, including technical, organizational and operational elements shall be included in the risk assessment.

The OGP risk matrix consists of a consequence axis and a likelihood axis. The consequences are those of credible scenarios (taking the prevailing circumstances into consideration) that can develop from the release of a hazard. The potential worst case consequences, rather than the actual ones (that may have occurred

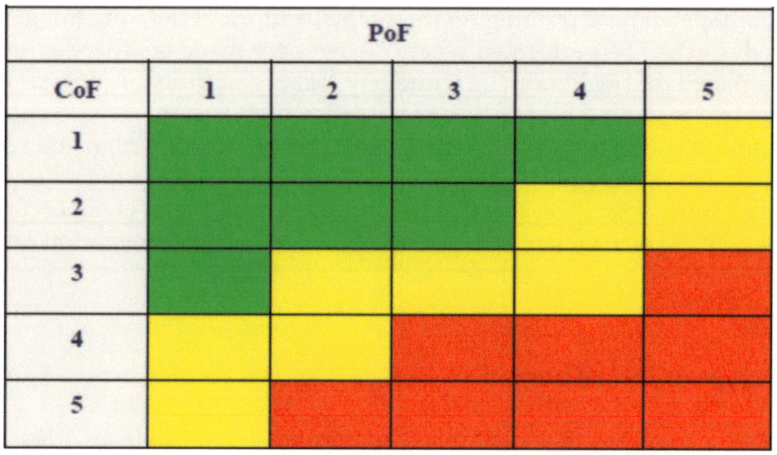

Figure 8.
Typical risk matrix used by oil and gas producers (OGPs).

previously), are used. After assessing the potential outcome, the likelihood on the vertical axis is determined on the basis of historical evidence or experience that such consequences have materialized within the industry, the entity or a smaller unit (**Figure 8**).

3.4 Maintenance management system

Effective inspection and maintenance are important in ensuring asset integrity and reliability. In developing the maintenance management systems an initial review is required determine status and how the aging processes is covered in the existing maintenance program. The review is to evaluate the need for updating the integrity, reliability, vulnerability and consequence analysis for continuous operations in the future. Experience and knowledge from documented failures and lessons learnt shall also be part of the evaluation and be used to improve the maintenance management system. In principle, the maintenance management system should be within a computerized database with detailed history of the operating, design, assessment, inspection and maintenance records accessible to all key personnel.

4. Emergency preparedness

A review of the current emergency response systems must be performed, including an evaluation of how operational changes and new requirements will be met in the period of life extension. If there is a change in operating philosophy, HSE Case shall be revisited. OGPs are to evaluate any likely operational or organizational changes to the facilities that will affect the emergency preparedness and response systems.

5. Organizational and human factors

Human factors area comprises methods and knowledge which can be used to assess and improve the interaction between people, technology and organization to realize efficient and safe operations. The factors should include organizational structure, competency or training requirements, and succession planning. Human factors analysis shall be performed where changes are made or where extended life challenges the established human, technology and organizational context. Organizational system is also a factor to be considered, which aspects include engineering design, contract and procurement management. Engineering design and related procurement activities require a thorough and careful consideration of asset aging and life extension factor. The risk from each finding and the overall potential (future) risks shall be evaluated before deciding on the implementation of measures.

6. Assurance and verification

The OGP is to ensure that experience on lifetime extension from other installations and operating areas is applied to the analyses and evaluations carried out for the application. Any specific relevant information shall be included in the application document. OGPs are to ensure that the analyses and evaluation work has been carried out in accordance with the regulations, the relevant company standards and have been verified by the appropriate technical discipline authority.

7. Occupational health

The OGP will evaluate the status of working environment factors that are relevant for life time extension, prior to the commencement of implementing ALE. Factors that should be include considerations for chemical/radiation exposure, lightning and ventilation, ergonomics, noise/vibration pollution, material handling and storage, outdoor operations and accommodation facilities.

The main objective of the evaluation is to provide a status of the working environment according to both technical and operational requirements. The assessment/evaluation are appropriately based upon existing conditions at the facility, and if necessary, follow-up with new evaluations and assessments as required. The operational risks of each from each finding and the future risks shall be evaluated before deciding on the implementation of measures for improving working environment.

8. Engineering design

All assets are required to have design documentation available and accessible, which supports effective design at all stages of the asset life cycle and in relation to the management of aging life extension. All engineering activity to be undertaken throughout the anticipated service life of an asset should properly address life extension considerations.

9. Asset life extension for fixed offshore structures

Zettlemoyer [8] provides a template for the asset life of fixed offshore structures. In general, the main source of interest or ALE involves the jacket substructure which are essentially made up of tubular steel sections welded together to form a truss system. For fixed offshore structures, the jacket template or truss system is considered as a structural safety critical element (SCE), so the integrity must always intact (**Figure 9**).

The jacket template substructure is to be assessed for ALE in terms of ultimate strength of the structure and fatigue life assessments. The values for the ultimate strength results are represented in terms of an RSR (Reserve Strength Ratio). The RSR is the ratio of the base shear at collapse/base shear at the 100 Yr environmental loading (i.e. the design condition). Ultimate strength analysis is also called a pushover analysis or collapse analysis and involves non-linear analytical methods (**Figure 10**).

For new structures a RSR value is generally over 2.0. As platforms operate for some time, degradation due to corrosion, damage due to accidental damage (vessel collision, dropped objects) is possible. Offshore structures are inspected and anomaly management is performed to detect and repair damage based on severity levels over its operations. It is expected that the RSR may be compromised and reduced if damage is unmitigated.

For every operating region the acceptance criteria need to be determined as it varies from region to region for varying levels of environmental loading. In the Gulf of Mexico (GoM) minimum acceptance criteria based on API RP2A are provided in **Table 1**.

The operations for fixed offshore structures can be extended if the platform in principle has greater than its minimum RSR values. For asset life extension, it is expected that all severe damage to structures has either been repaired or reduced to a manageable condition, prior to the migration to ALE. In essence the topsides of the fixed offshore needs to be appropriately assessed. This is generally done by using a risk based maintenance (RBM) program where anomalies are rectified due to severity levels. These topside RBM should be aligned to other topsides programs including piping, equipment skids and vessels to ensure that the maximum use can be made of allocated resources.

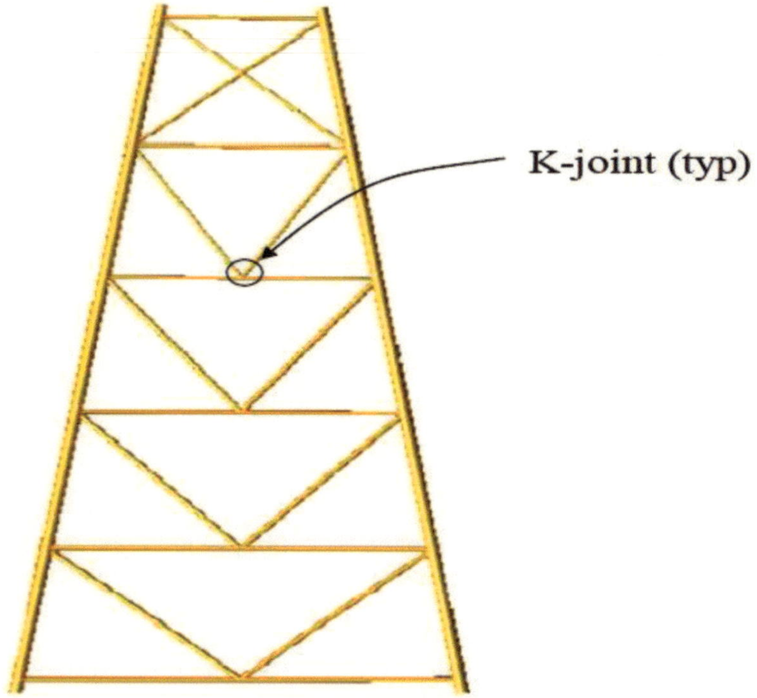

Figure 9.
Typical jacket substructure for a fixed offshore structure [8].

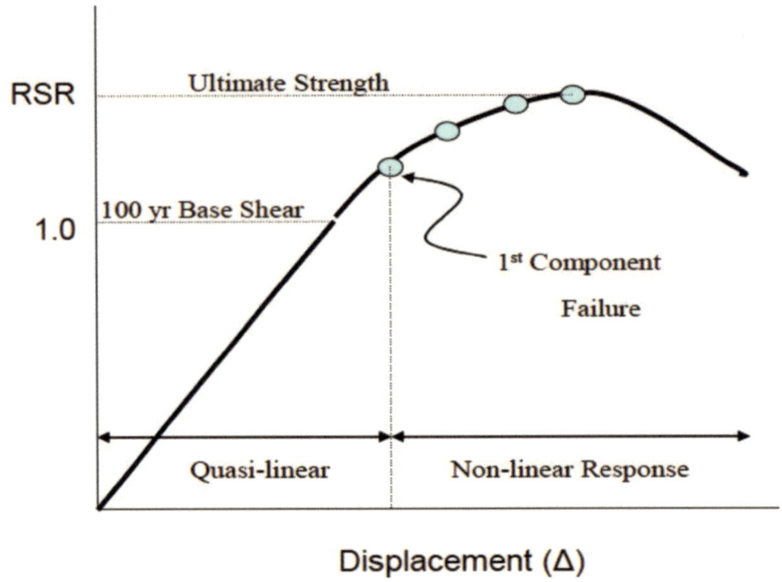

Figure 10.
Typical schematic for load displacement curve—ultimate strength [8].

Assessment Category	Minimum RSR	
	GoM	West Coast
A-1	1.2	1.6
A-2	0.8	
A-3	0.6	

Table 1.
Minimum RSR values for the Gulf of Mexico (API RP 2A, 2014) [9].

The other key performance indicator for tubular joints is fatigue life prediction. In recent years through the use of proper joint configuration in design, use of joint flexibility approaches in analysis, the issue of fatigue life estimation has been argued away. In principle the acceptance criteria for ALE for a fixed offshore structure is the exceedance of the minimum RSR values. In recent years, integrity management codes of practice including API RP2SIM (2014) [10], ISO 19902:2007 [11] provide guidance on the ALE for fixed operating steel structures.

10. Case study

10.1 Aging electrical component

Facilities are designed typically with a life span of 20–25 years. However, it is becoming common for facilities, both on-shore and off-shore, to be operated beyond its life span. While assets are designed for 20–25 years, equipment age differently and suffer from different age-related failure mechanisms. Other than aging, electrical, control and instrumentation equipment suffer from obsolescence. This is primarily due to unavailability of components and end of hardware/software support.

10.2 Situation

A downstream refinery experienced control issues with its two reactor and regenerator slide valves on its Residue Fluid Catalytic Cracker (RFCC) unit. The symptom, initially, manifested as valve hunting. This progressively worsen to the point where the valves had to be put on manual hand wheel control. While this action temporarily stopped the valve hunting, it made control of the reactor and regenerator catalyst level very difficult as operators had to be on-site to adjust the valve opening manually. Left unresolved, the likely consequence of this situation was a process upset and RFCC unit trip. This would also cause a cascade effect, resulting in the shutdown of other units, incurring significant production losses and HSE exposure (**Figure 11**).

10.3 Problem analysis

The problem was initially thought to be due to a failure of the HPU control module. However, the problem was traced, eventually, to a failing DC power supply unit (PSU) which powers the control module. Once the problem was identified, replacement of the power supply unit resolved the issue. What should be noted from this seemingly straightforward problem is that both reactor and regenerator HPU units had suffered from the same problem. On closer inspection, both power supply units were of the same make and had been first installed (as part of the HPU units) at around the same time. At the time of the incident, the power supply units

Figure 11.
A typical FCCU slide valve actuator (left) and HPU & control module (right) [12].

were estimated to be 10 years old. Aging was attributed as the cause of the problem, as within 6 months, another slide valve HPU had also suffered from an almost similar problem.

There are several failure mechanisms that are typically found due to aging. Unfortunately, a detail inspection of the power supply unit was not carried out to identify the aging mechanism. **Table 2** shows common aging mechanism for primary containment (piping, vessels, heat exchangers), structures, safeguarding systems and electrical, control and instrumentation (EC&I) [HSE UK RR823 Plant Aging Study] [8].

10.4 Solution

The multiple failures within a short period of time was a strong indication of an age related failure, as opposed to a random failure. As a result, several actions were taken:

- All HPUs with PSU of similar make and type were identified.
- PSUs were replaced (like-for-like replacement).
- A Preventive Maintenance (PM) plan was created in PMMS to replace the PSUs every 8 years.
- Learnings (failure mode, failure mechanism, failure rectification) were incorporated into the technician training program for future ease of troubleshooting.

Due to the potentially high consequence of production loss from this failure, other components of the HPU were also scrutinized. Other critical components, and possible single points of failure, were identified. These were parked for future improvements for the next asset refresh cycle. A similar exercise to this would be to perform a failure mode, effects and criticality analysis.

To ensure these upgrades were implemented in the next possible opportunity, the equipment upgrade was parked into the refinery's 5-year CAPEX plan. The site's

Ageing Mechanism	Primary Containment	Structures	Safeguards	EC&I
Corrosion	✓	✓	✓	✓
Stress Corrosion Cracking	✓	✓	✓	
Erosion	✓	✓	✓	✓
Fatigue	✓	✓	✓	
Embrittlement	✓	✓		✓
Physical damage	✓	✓	✓	✓
Spalling		✓		
Weathering		✓	✓	
Expansion/ contraction due to temperature changes (process or ambient) or freezing	✓	✓	✓	✓
Instrument drift				✓
Dry joint development				✓
Detector poisoning				✓
Subsidence		✓	✓	

Table 2.
Typical aging damage mechanism (HSE UK RR823 Plant Aging Study) [13].

Equipment Obsolescence Masterplan was also updated. This is typically reviewed on a yearly basis to manage overall life cycle of aging EC&I assets. **Figure 12** is an example of an equipment obsolescence dashboard which lists EC&I equipment asset, obsolescence status and remedial plan.

This case study highlights several important aspects of managing aging assets:

- Information is in the data. Useful insights can be obtained through data analysis. Equipment failure rate, for example, will show whether an equipment is approaching end-of-life. However, quality data is essential and data clean-up often is required before analysis can be done.

- Obsolescence management is essential for EC&I equipment. All equipment should be captured in an asset list and the aging strategy should be clearly defined. This could be through various way which include replacement, upgrade, life extension (through supplier extended support), life extension (with available spares) or run-to-fail.

- EC&I equipment typically will have shorter life-cycle than an asset overall design life. Therefore, EC&I aging strategy has to be put in place much earlier than other assets such as structures and mechanical equipment. With E&CI equipment, analysis down to the major component level (e.g. PSU) should be done. This may need to also include supporting equipment such as interface modules, equipment communicators and workstations as well as software.

- Remediation of aging asset can be based equipment criticality and/or actual equipment condition. There are various methodologies that can be employed to determine equipment criticality such as failure mode, effects and criticality

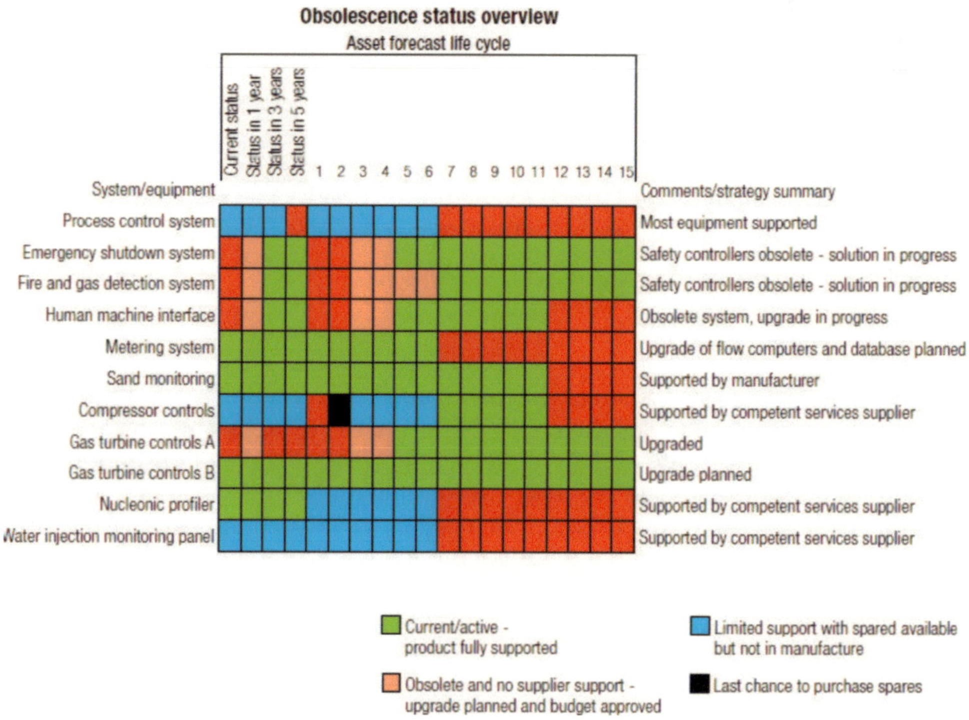

Figure 12.
Example of an equipment obsolescence dashboard (HSE UK KP4 report) [7].

analysis (FMECA), reliability availability and maintainability (RAM) analysis, and even layers of protection analysis (LOPA), among others. Each methodology puts a different focus on equipment reliability and integrity.

- The human aspect of managing aging asset should not be under-estimated. Knowledge is lost when people move or retire. Therefore, knowledge retention is key in ensuring assets can continue to be managed safely and reliably.

11. ALE reporting requirements

As a minimum, OGPs should establish the following in their submission of ALE Study Consent for Extension Report:

- Clarity on how the asset is to be operated during the extension period.
- Clarity on Fitness for Service to run up to Design Life, Remnant Life Assessment, and Life Extension requirement and Gap Closure requirement for the Asset
- Economic Analysis is performed with the following scenarios:

  1. No Further Production Enhancement Action, for three (3) different options of Crude Oil Price.
  2. Shortest Extension Period, for different options of Crude Oil Price.
  3. Longest Extension Period based on the longest remnant life of a discipline assessed, for different options of Crude Oil Price.
  4. Three further scenarios of extension period in between shortest and longest period for different options of Crude Oil Price.
  5. Sensitivity analysis for Capital Expenditure (CAPEX) and Operational Expenditure (OPEX) for a variety of scenarios.

12. Conclusions

Oil and gas producers are often driven to continue operations beyond its design facility and are required to operate safely. There are many factors to consider when providing an asset life extension solution to aging offshore or onshore facilities. This chapter presents the key issues to consider and a prescribed methodology follow in justifying asset life extension for an aging facility. At all stages of the asset life extension, assets are required to satisfy the As Low as Reasonably Practicable criteria as a minimum for each discipline and demonstrate fitness for purpose.

Acknowledgements

The authors of this chapter will like to thank the PETRONAS Operations Excellence, Centre of Excellence management team for their kind support of this project.

Author details

Riaz Khan[1*], Ammeran B. Mad[1], Khairil Osman[2] and Mohd Asyraf Abd Aziz[2]

1 PETRONAS Sdn Bhd, Kuala Lumpur, Malaysia

2 Pure Integrity Sdn Bhd, Kuala Lumpur, Malaysia

*Address all correspondence to: riaz.khan@petronas.com

References

[1] Oil Rig/Offshore Structure [Internet]. 2009. Available from: http://oilandgasprocessing.blogspot.com/2009/02/oil-rig-offhore-structure.html

[2] Vopak Tank Storage Terminals [Internet]. 2019. Available from: https://www.vopak.com/terminals/asia/vopak-singapore-sebarok-terminal

[3] Corrosion, materials and integrity [Internet]. 2018. Available from: https://www.gateinc.com/oil-gas-engineering-services/materials-corrosion/

[4] ESReDA. Ageing of components and systems. In: Petersson L, Simola K, editors. An ESReDA Working Group Report. 2006. pp. 112-127

[5] 122 Norwegian Oil and Gas Recommended Assessment and Documentation for Service Life Extension of Facilities, Rev1. 2012

[6] Hokstad P, Habbrekke S, Johnsen R, Sanglesland S. Ageing and Life Extension for Offshore Facilites in General and for Specific Systems. SINTEF Society and Technology, Norsk olje & gass; 2010

[7] Key Programme 4 (KP4) Ageing Life Extension Programme. A Report by the Energy Division of HSEs Hazardous Installation Directorate. UK: Health & Safety Executive; 2014

[8] Nicolas Z. Life extension of fixed platforms. In: Young B, editor. Tubular Structures XIII. London: CRC Press; 2010. DOI: 10.1201/b10564-3

[9] Planning, Designing and Constructing for Fixed Offshore Structures. Recommended Practice. APIRP 2A 22nd ed. American Petroleum Institute; 2014

[10] Structural Integrity Management for Fixed Offshore Structures. Recommended Practice. American Petroleum Institute, APIRP2SIM. 2014

[11] Petroleum and Natural Gas Industries-Fixed Offshore Structures. Recommended Practice. American Petroleum Institute, British Standard, ISO 19902; 2007

[12] FCCU Valves and Control Systems. [Internet]. 2018. Available from: http://www.tapcoenpro.com/

[13] RR823 Plant Ageing Study. Phase 1 Report. Prepared by ESR Technology Limited. UK: HSE; 2010

Chapter 6

Impact of Improving Machines' Availability Using Stochastic Petri Nets on the Overall Equipment Effectiveness

Aman Zineb, Latifa Ezzine and Haj El Moussami

Abstract

The objective of this chapter is to demonstrate the robustness of stochastic petri nets in the field of maintenance for the improvement of machine availability. We aim to present the modeling of the maintenance function in a production site with stochastic Petri nets by using two performance indicators: the mean time between failures (MTBF: the time between two successive failures) and the mean time to repair (MTTR: average time to repair) to improve the equipment performance. The determination of the distribution law is essential for each statistical study and provides a powerful and reliable model for the evaluation of the equipment performance. After determining these laws we switched to modeling Petri nets, we proposed the establishment of an effective preventive maintenance plan which aims at increasing the reliability, thus reducing the probability of failures. Consequently, we increase the machines' availability and then the overall equipment effectiveness.

Keywords: maintenance, modeling, stochastic petri nets, performance indicators, availability, overall equipment effectiveness

1. Introduction

Nowadays, companies are in total competition aiming the development of new industry strategies especially concerning the maintenance and the production planning. In fact, general customer non satisfaction is noted to be a consequence either of random demand or sudden equipment failure. Therefore, it becomes a necessity to develop new maintenance and production strategies.

The maintenance function is largely considered as a non productive one since it does not afford us with currency directly. In fact, each company should produce goods of a certain quality and quantity required by the customer who needs to receive the product at the right time. To achieve this objective, production systems should operate efficiently and accurately by using an effective maintenance plan helping companies maximizing availability by minimizing machine downtime. Our work is the continuity of the previous one that manifested the obligation of improving availability [1, 2]. The machine suffered several failures. That's why we want to find a law that will helps us anticipate future failures and when these failures will occur. The aim of this paper is to propose a modeling and performance evaluation method for a production site using stochastic Petri nets as a very powerful modeling tool that contributes to the improvement of the availability.

In order to exhibit the role played by the maintenance function within the company, it seems important to focus first on the huge difference between planned downtime and unplanned downtime. Unplanned downtime should be minimized. It is function of the number of breakdowns within a specified time period and related measures such as mean time between failures (MTBF) and mean time to repair (MTTR) [3]. MTBF and MTTR are claimed to be measures of equipment achievement and are related to objectives such as functional performance and process capability [4]. Thorough analysis of these two elements enables the maintenance function to improve equipment's availability by either increasing the MTBF or reducing the MTTR.

In this chapter, we show the interest of stochastic Petri nets for modeling, evaluation and performance analysis. Therefore, we will present two applications of SPNs to generate the model after having conducted a statistical study to determine transitions laws. Based on the model and after its simulation, we calculated the availability of machines after having developed a high-quality preventive maintenance plan allowing us to notice the impact of the increase value of availability on OEE.

2. Literature review

Petri net models have been studied extensively over the last decade [5–8]. These models have been applied to many types of systems [9–16]. A lot of analysis has been made on different states a system may occupy. These analyses had not take into consideration any study of timing. Recently, timing is integrated in some attempts [14, 17]. Merlin and Farber [18] discussed timed Petri nets. In fact, they assigned a time threshold and maximum delay to a transition. This was done to allow the incorporation of timeouts into a protocol model.

In his work, Zuberek [17] presents a fixed time for each transition to model the performance of a computer system at the register level. In another case, probability was introduced to allow a random switching of flow through the graph [19]. Shapiro limits the model to discrete time and a maximum of one token in each place. In this paper, Petri nets are extended by assigning an exponentially distributed firing rate to each transition for continuous time systems or a geometrically distributed firing rate to each transition for discrete time systems. These new stochastic Petri nets (SPNs) are isomorphic to homogeneous Markov processes [20]. In this work, SPN's are used to model the maintenance field in order to increase availability, productivity and efficiency of the production line.

Many stochastic Petri Nets classes are proposed for performance analysis of production systems. The characteristics of the different classes are essentially in the nature of transitions used, where laws other than exponential are associated [20–22].

To model our system we have used the stochastic Petri Nets with predicates because they consist of both, the immediate transitions, the deterministic transitions and the transitions with stochastic timings distributed with any law. In addition, they use variables to include two other properties:

- "Guards": variables or Boolean expressions that make the transitions uncontrollable until they are verified.

- "Assignments": assignments that modify the values of variables when crossing transitions.

To define a maintenance policy, prioritize the interventions or establish the budget, the maintenance manager must be able to choose the means and modes of intervention most suited to his machines. Similarly, a working group aimed at making machines, a line, a cell or a workshop more reliable, requires a structured method of attacking the site. Here are some simple indicators to help with the decision.

2.1 Reliability index: mean time between failures (MTBF)

Mean time between failure (MTBF) indicator measures the time elapsed between failures. It is therefore beneficial for reliability and a fortiori for availability. Mathematically, the criterion of reliability is thus defined by the inverse of its indicator. MTBF is the mean time between consecutive failures. For a repairable system, the MTBF is the average time between completion of the repair and the next failure and is calculated using (1).

$$\text{Mean time between failures} = \frac{\text{total up time}}{\text{number of breakdowns}} \tag{1}$$

Total up time includes stopping time off failure and micro stops time.

2.2 Maintainability index: mean time to repair (MTTR)

Maintainability means, for an entity used under given conditions, the likelihood that a given maintenance operation can be performed over a given time interval, maintenance is provided under given conditions, and use procedures and means. MTTR is calculated using (2).

$$\text{Mean time To repair} = \frac{\text{total down time}}{\text{number of breakdowns}} \tag{2}$$

2.3 Rate of availability

The notion of availability expresses the probability that an entity is in a state of "availability" under given conditions at a given time, assuming that the provision of external means is assured. This rate is calculated using (3).

$$\text{availability} = \frac{\text{MTBF}}{(\text{MTBF} + \text{MTTR})} \tag{3}$$

3. Applications

In this section, we present two applications in automotive and food sector where stochastic Petri Nets proves its strength in order to increase equipment's availability.

3.1 Electrical test table (automotive sector)

In this first application, we present a study to improve availability of equipment that guarantees the good quality of the wiring harnesses. It is the electrical test table whose role is to check the flow of the electric current and therefore it ensures the main technical function of the wiring harnesses produced.

To achieve this result, we present the electrical test table and its components. Then, availability is calculated based on the history provided by the company. After this, we find the law that will help us anticipate future failures and when these failures will occur. This law enables us to develop a high-quality preventive maintenance plan in order to increase the availability of the equipment.

3.1.1 Description and decomposition of electrical test table

It is a matter of clearly identifying the elements of the machines to be studied in order to analyze, for each element, the risks of dysfunction. We proceed with a general analysis followed by operational analysis. It is first of all necessary to formulate the need in the form of simple functions that the equipment must fulfill by answering three basic questions shown on the standardized tool called the horned beast diagram illustrated in **Figure 1**.

Operational analysis is the analysis of the decomposition of the electric test table; this analysis will be made by a simple decomposition into blocks of element presented in **Figure 2**.

3.1.2 Determination of the transition laws

After describing the electric test table to be studied, we are confronted with the problem of determining the failure and repair function of this equipment.

In order to model the breakdown/repair process of a repairable system we have used the history of the operating time of this equipment. The failure history of the electric test table machine enabled us to analyze the data and calculate the availability of this machine along the past period. After any calculation made, the availability of the machine is found to be of the order of 62.73%.

The choice of a particular model is statistically tested to select the model best suited to the observed failure and repair times. The protocol to be used for the study of repairable equipment has been developed by Ascher and Feingold [23].

Using the Anderson-Darling statistic and its p-value, we make the decision that the model, the exponential, the Weibull or Log normal, is the one that best adjusts the data. The Anderson-Darling test [24] is used to check if a sample of the data comes from a population with a specific distribution.

Minitab computes the Anderson-Darling statistic using the weighted quadratic distance between the fit line of the probability diagram (based on the chosen law and using the maximum likelihood estimation method or estimates of the least

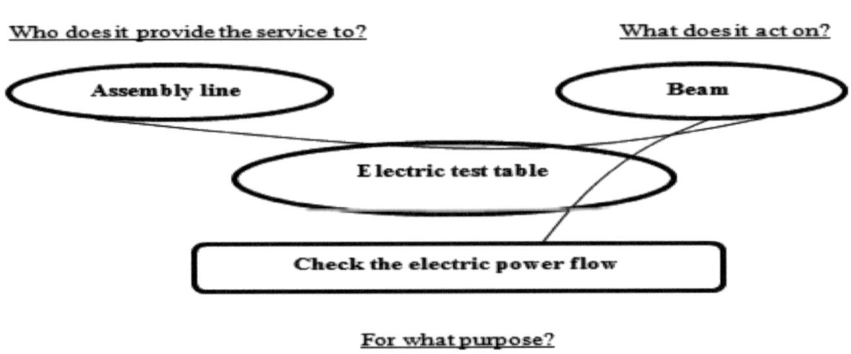

Figure 1.
Horned beast for the electric test table.

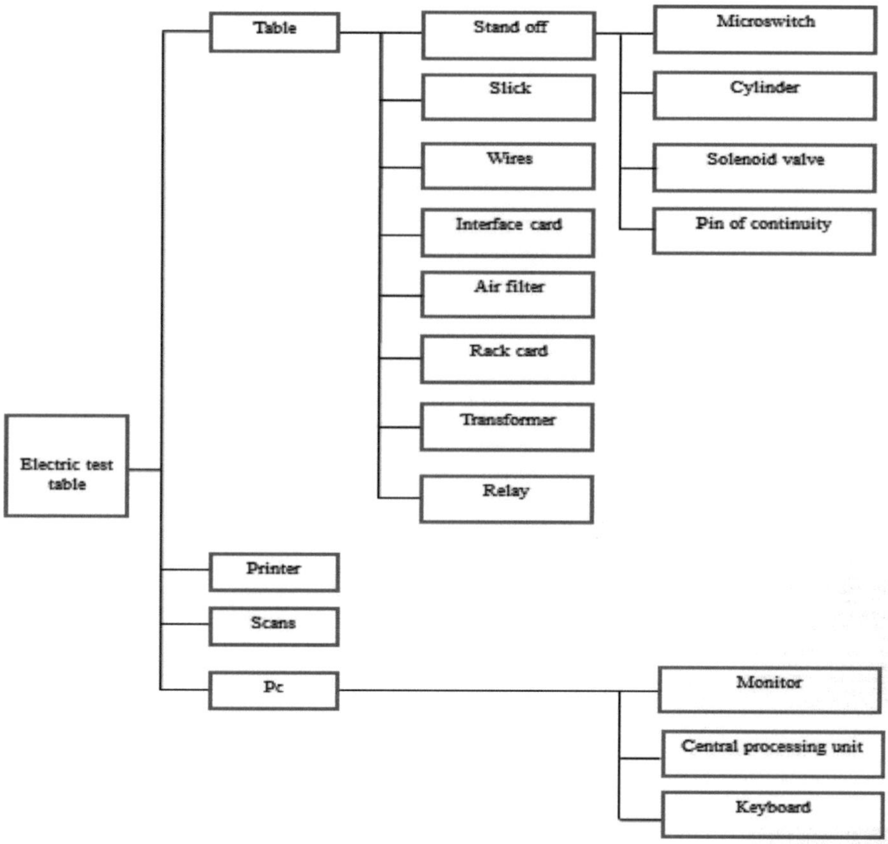

Figure 2.
Decomposition of electric test table.

squares) and the non-parametric staircase function. The calculation is weighted more extensively at the ends of the distribution.

The hypothesis test is defined as:

- H_0: the data come from a specific distribution.
- H_1: the data does not come from a specific distribution.

The decision to accept or reject the null hypothesis concerns the value p. If the value p is greater than 0.05, the null hypothesis is accepted, and if the value p is less than 0.05, the hypothesis is rejected.

3.1.3 Mean time between failure modeling

On the basis of the history of operating times, we used the data processing software MINITAB17 [25] in order to obtain the results mentioned in **Figure 3**.

According to the results obtained and by comparing the different values of P, we note that the largest values of p are:

0.522: This corresponds to the law: normal distribution with BOX-Cox transformation.

The Box-Cox transformation is a power transformation, W = Y*λ, in which Minitab determines the best value for λ.

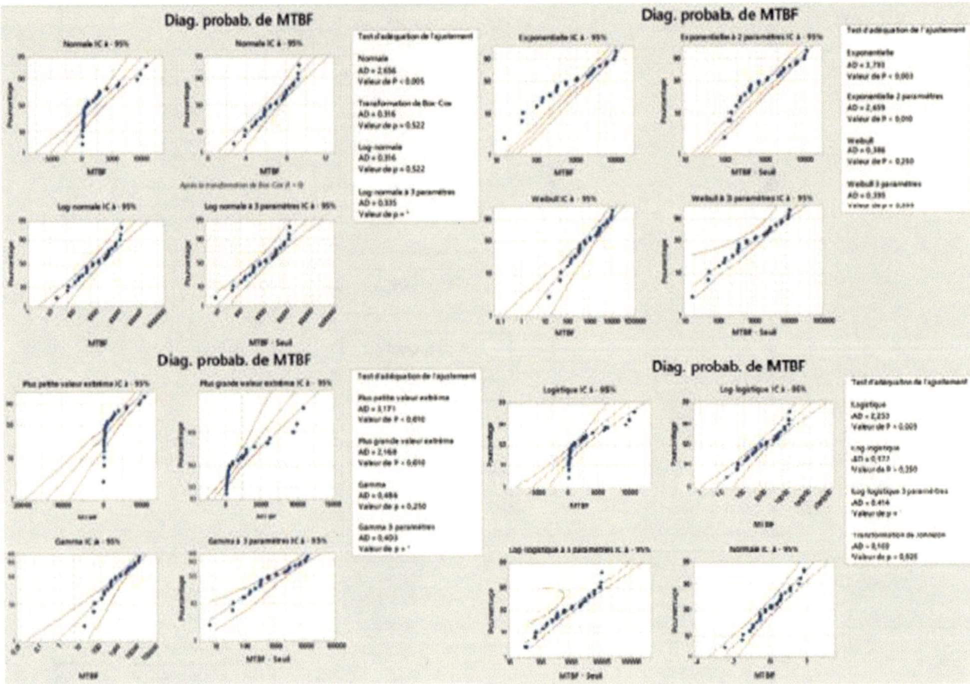

Figure 3.
Distribution laws for MTBF.

0.925: This corresponds to the law: normal distribution with Johnson transformation.

After the application of a Johnson transformation, the data closely follow a normal distribution; indeed, the value of p is high and virtually all data points lie within the confidence limits of the Henry line.

Box-Cox transformation: $\lambda = 0$

Johnson transformation: $1.15528 + 0.453754 \times \mathrm{Ln}((X - 10.5919)/(13309.4 - X))$

3.1.4 Mean time to repair modeling

Based on the history of the same equipment for the same time period, and using the MINITAB17 data processing software in order to obtain the results mentioned in **Figure 4**.

We note that the largest values of P are:

0.55: This corresponds to the law: normal law with transformation of BOX-Cox.

The Box-Cox transformation is a power transformation, $W = Y \times \lambda$, in which Minitab determines the best value for λ.

0.145: This corresponds to the law: normal law with Johnson transformation.

After the application of a Johnson transformation, the data closely follow a normal distribution; indeed, the value of p is high and virtually all data points lie within the confidence limits of the Henry line.

Box-Cox transformation: $\lambda = 1$

Johnson transformation: $-6.98629\mathrm{E}{-16} + 0.898955 \times \mathrm{Ln}((X - 1.44460)/(11.5554 - X))$

3.1.5 Stochastic Petri Net modeling for preventive maintenance plan

After the analysis and calculations, the following stochastic Petri net was realized in **Figure 5**.

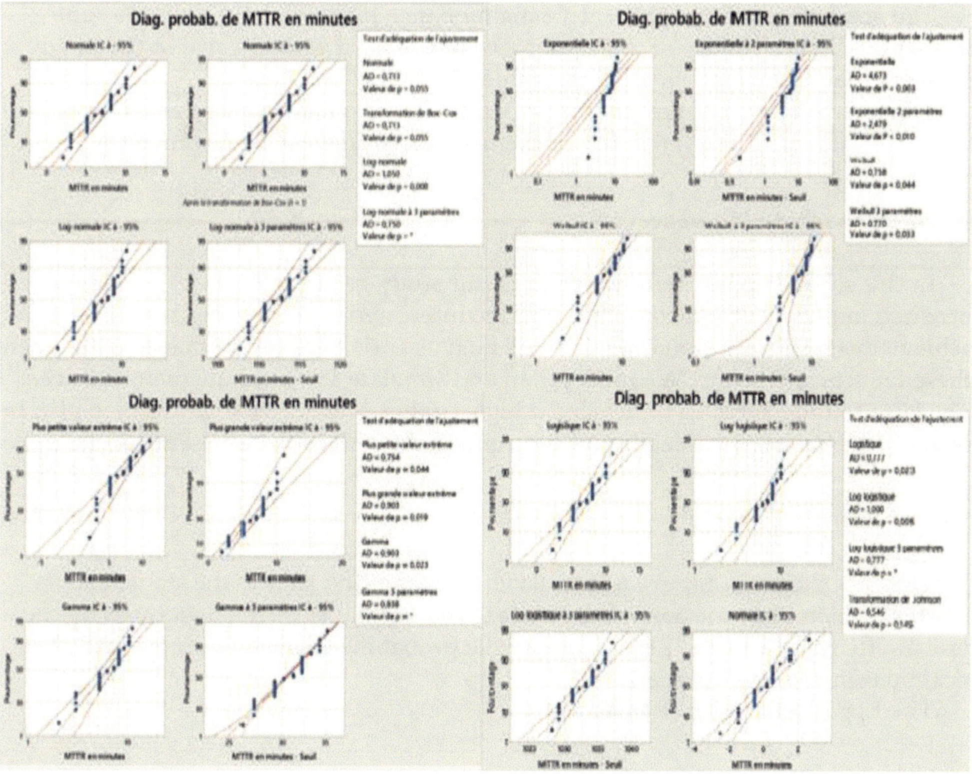

Figure 4.
Distribution laws for MTTR.

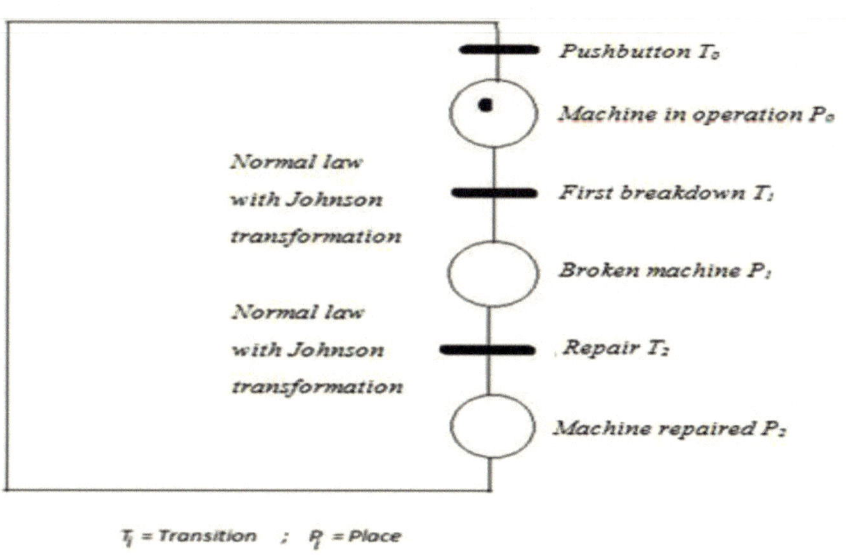

Figure 5.
Stochastic petri net modeling.

After modeling the maintenance function of the electric test table, it became easy to predict the next breakdown and when it will occur. So, we developed a high-quality preventive maintenance plan based on the modeling realized.

The application of this preventive maintenance plan will have an important impact on the availability of the equipment which is best manifested in the progress of the availability which reached the value of 97.05%.

Thus, the strength of the stochastic petri nets was demonstrated as a tool of modeling allowing the availability of the machine studied to be improved.

3.2 Sieve machine (food sector)

In this second application, we present our study to improve the efficiency of a production line in a food company. We are interested in improving availability. To achieve this result, we find laws that will help us anticipate future failures and when these failures will occur. We model then and simulate the machine maintenance. The laws found enable us to develop a high-quality preventive maintenance plan in order to increase the availability of the equipment using stochastic Petri Nets.

3.2.1 The normality test

This test shown in **Figure 6** is considered as the first step of the statistical mastery which makes it possible to analyze the normality of the data by using the probability plot (p-value*) that is to say the probability that two samples are identical by using a test hypothesis.

The hypothesis test is defined as:

- H_0: the data follow the normal law;
- H_1: the data do not follow the normal law.

The decision to accept or reject the null hypothesis concerns the value p. If the value p is greater than 0.05, we accept the null hypothesis, and if the value p is less than 0.05, we reject the hypothesis.

For the TTR we found p-value = 0.335 > 0.05. So we will accept the hypothesis H_0 and we can say that the data of the TTR follow the normal law.

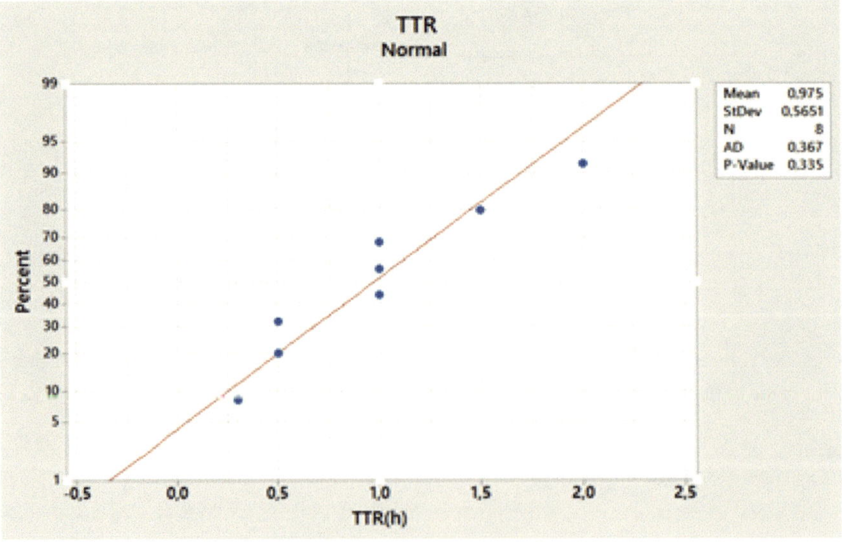

Figure 6.
Normality test for TTR.

3.2.1.1 The tests of the laws on Minitab 17

Minitab proposes Anderson-Darling Statistics and its p-value to make the decision on which model the data is distributed in. The Anderson-Darling test [24] is used to test whether a sample of the data comes from a population with a specific distribution.

The hypothesis test is defined as:

- H_0: the data come from a specific distribution;
- H_1: The data does not come from a specific distribution.

The decision to accept or reject the null hypothesis concerns the value p. If the value p is greater than 0.05, we accept the null hypothesis, and if the value p is less than 0.05, we reject the hypothesis.

On the basis of the history of operating times, we used the data processing software MINITAB17 [25] in order to obtain the results mentioned in **Figure 7** for TTR.

From the results obtained we notice that the value of p-value of the TTR for all distributions is greater than 0.05. This last value allows us to accept the hypotheses H_0.

3.2.1.2 Adjustment tests on XL-Stat

Using this software, we performed a law fit with a risk of 5% and we adopted the method of estimation of maximum likelihood.

The different laws tested and their value of P can be summarized in **Table 1**.

The distribution that best fits the data for the fit test is the Weibull distribution (2).

Its estimated parameters are grouped in **Table 2**.

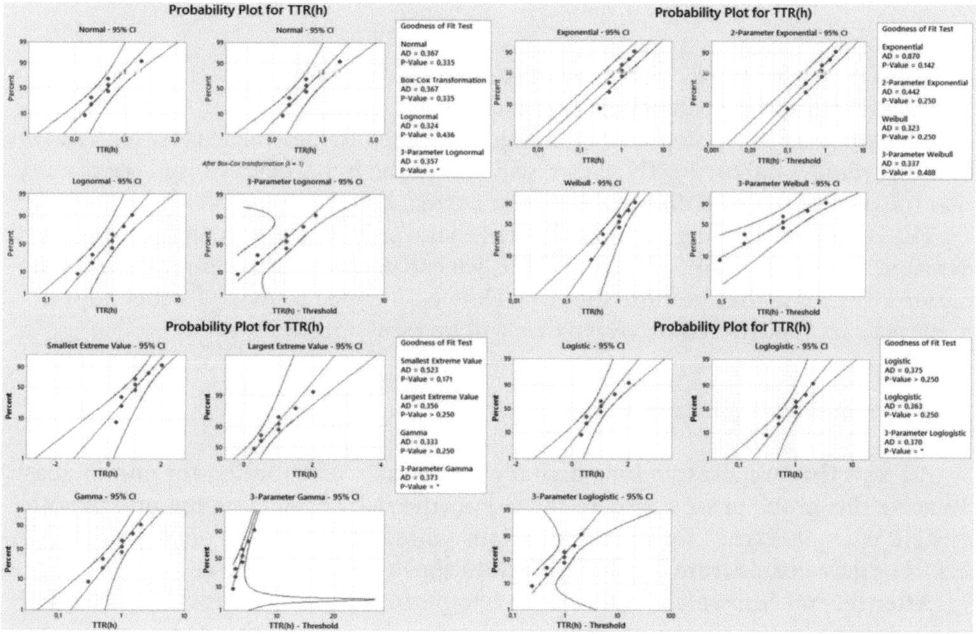

Figure 7.
Anderson-Darling analysis of the TTR.

Distribution	p-Value
Normal standard	0.010
Student	0.010
Fisher-Tippett (1)	0.064
Gumbel	0.089
Gamma (1)	0.191
Erlang	0.200
Khi^2	0.516
Weibull (1)	0.560
Exponential	0.638
Log-normal	0.774
Gamma (2)	0.901
Normal	0.929
Logistic	0.964
GEV	0.966
Weibull (2)	0.969

Table 1.
The TTR distributions and their p-value.

Parameters	Value	Standard error
Beta	1.919	0.659
Gamma	1.187	0.267

Table 2.
The estimated parameters of Weibull distribution (2).

The XL-stat software offers several tests to ensure this distribution: Kolmogorov–Smirnov test and chi-square test.

Following the same steps, after having conducted the normality test for TBF, we found p-value = 0.020 < 0.05. So we will accept the hypothesis H_1 and we can say that the data of the TBF do not follow the normal law.

The Anderson-Darling analysis showed that, for TBF, the value of p for some distributions is less than 0.05. As a result, we can conclude that none of the distributions is the one that best fits the data. Hence, the need to make further adjustment tests on the XL-Stat software that will be most appropriate.

3.2.2 Determination of transition laws

All tests that are already done do not give an exact distribution for our database. To solve this problem we compare the risk α (the risk of rejecting the null hypothesis H0 when it is true) for the distributions whose value of p-value is greater than 0.5. (H_0: data comes from a specific distribution) (**Tables 3** and **4**).

After several adjustment tests on the two performance indicators TTR and TBF and using the comparison between the risk values α, we find that the good distribution for the TTR is the Exponential distribution and for the TBF is a normal Log distribution.

Distribution	p-Value	α (%)
Weibull (1)	0.560	32.43
Exponential	0.638	10.65
Log-normal	0.774	21.40
Gamma (2)	0.901	28.72
Normal	0.929	27.37
Logistic	0.964	22.06
GEV	0.966	20.60
Weibull (2)	0.969	32.43

Table 3.
The TTR distributions and their risk α.

Distribution	p-Value	α (%)
Exponential	0.382	80.4
Fisher-Tippett (2)	0.593	1.13
Normal	0.642	0.37
Logistic	0.759	0.25
Log-normal	0.800	0.5
Gamma (2)	0.882	7.31
Weibull (2)	0.906	4.40

Table 4.
The TBF distributions and their risk α.

Performance indicator	Distribution law	Parameters	Value
TTR	Exponential	λ_1	0.952
TBF	Normal Log	μ	2.097
		σ	1.726

Table 5.
Parameters estimated for each law.

According to the method of maximum likelihood, we estimate the value of the parameter of each law (**Table 5**).

3.3 Petri net modeling for preventive maintenance

For our case, we worked on modeling a single machine (Sieve) using stochastic Petri nets because we can use this type of model to take into account probabilistic events such as the failure of a machine moreover it allows to model tasks with non deterministic execution times and to evaluate the performances of the system.

The Petri Net modeling in **Figure 8** represents the behavior of a sieve.

This network has two places:

- Pl1: Running
- Pl2: Out of order

And two transitions:

- Tr1: Equipment failure
- Tr2: Equipment repaired

The simulation results are presented in **Tables 6** and 7.

From the results of the simulation, it can be seen that the residence time of the sieve in working order is equal to 24.13 times the residence time in the state of failure. For the crossing frequencies of the two transitions, they are approximately equal in view of the nature of our operating process, that is to say a loop sequence.

After modeling the maintenance function of the sieve, it became easy to predict the next breakdown and when it will occur. So, we developed a high-quality preventive maintenance plan based on the modeling realized.

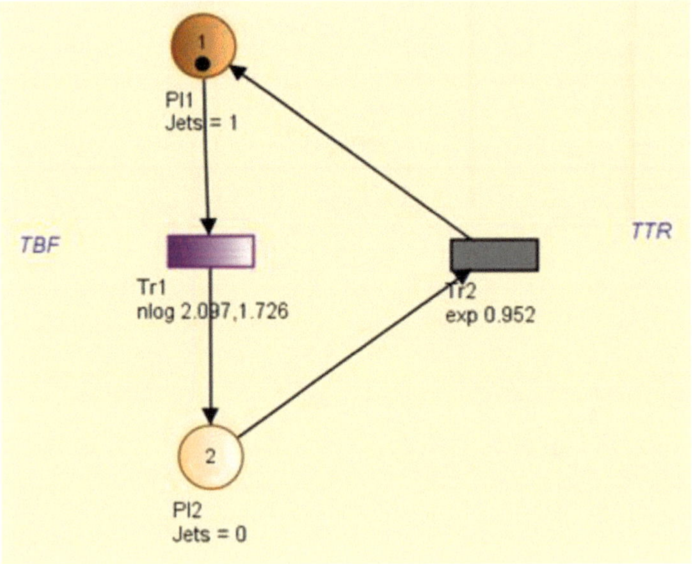

Figure 8.
Modeling with Petri nets.

Name	Number	Residence time	Standard deviation (σ)	Number of jets	Mean standard deviation	Number of jets	Final standard deviation
Pl1:1	1	7.1439E2	7.3264	0.9602	9.8473E−3	0.9	0.3162
Pl2:2	2	2.9612E1	7.3264	3.9801E-2	9.8473E−3	0.1	0.3162

Table 6.
Simulation result by GRIF-residence time of places.

Name	ID	Frequency of sorting period transitions
Tr1: 1	1	3.07E1
Tr2: 2	2	3.06E1

Table 7.
Simulation result by GRIF-Frequency of sorting period transitions.

The modeling allowed us to conclude that our machine has a good availability as the time of breakdowns is negligible compared to the time of stay in operation.

The application of this preventive maintenance plan will have an important impact on the availability of the equipment which is best manifested in the simulation conducted. According to the calculation of the performance using this modeling, we find an average availability of the sieve which equals:

$$D = \frac{714.39}{714.39 + 29.612} = 96\% \quad (4)$$

Thus, the strength of the stochastic petri nets was demonstrated as a tool of modeling allowing the availability of the machine studied to be improved.

3.4 Synthesis

In these applications, we have demonstrated the robustness of stochastic petri nets in the field of maintenance in two different sectors for the improvement of machines' availability.

Our study is based on the values of the two indicators (MTBF and MTTR) calculated within a company working in automotive sector and another one operating in food sector. The determination of the distribution law is essential for each statistical study and provides a powerful and reliable model for the evaluation of the equipment performance by incorporating preventive maintenance. After that, we developed a preventive maintenance plan that improves availability of machines.

For each sector, we obtained an increase in availability rate thanks to SPNs methodology.

4. Conclusion

The objective of this chapter is to demonstrate the robustness of stochastic petri nets in the field of maintenance for the improvement of machine availability. We presented the modeling of the maintenance function in a production site with stochastic Petri nets by using two performance indicators: the mean time between failures and the mean time to repair to improve the equipment performance.

The determination of the distribution law is essential for each statistical study and provides a powerful and reliable model for the evaluation of the equipment performance in the automotive sector and the food sector. After determining these laws we switched to modeling Petri nets, we proposed the establishment of an effective preventive maintenance plan which aims at increasing the reliability, thus reducing the probability of failures. Consequently, we increase the machines' availability and then the overall equipment effectiveness.

Author details

Aman Zineb[1]*, Latifa Ezzine[2] and Haj El Moussami[1]

1 Mechanics and Integrated Engineering, ENSAM School, Moulay Ismail University, Meknes, Morocco

2 Modeling, Control Systems and Telecommunications, Faculty of Sciences, Moulay Ismail University, Meknes, Morocco

*Address all correspondence to: zineb.aman@gmail.com

References

[1] Aman Z, Ezzine L, Fattah J, Lachhab A. Improving efficiency of a production line by using overall equipment effectiveness: A case study. In: Proceedings of the International Conference on Industrial Engineering and Operations Management, Rabat, Morocco, April 11–13. 2017. pp. 1048-1057. [Indexed in SCOPUS]

[2] Aman Z, Ezzine L, Fattah J, El Moussami H. Improving efficiency of a production line by using value stream mapping and overall equipment effectiveness: A case study. In: Proceedings of the International Conference on Industrial Engineering and System Management, Meknes, Morocco, May 17–18. 2017

[3] Pintelon L, Gelders L, Puyvelde FV. Maintenance Management. Leuven, Belgium: Acco; 2000

[4] Wilson A. Asset Maintenance Management—A Guide to Developing Strategy & Improving Performance, Conference Communication; 1999

[5] Peterson JL. Petri nets. ACM Computing Surveys. 1977;**9**:223-252

[6] Hack M. The equality problem for vector addition systems is undecidable. Theoretical Computer Science. 1976;**2**: 77-95

[7] Holt AW et al. Information System Theory Project Report, Tech. Rep. RADC-TR-68-305. NY: Rome Air Development Center, Griffiss Air Force Base; 1968

[8] Miller RF. A comparison of some theoretical models of parallel computation. IEEE Transactions on Computers. 1973;**C-22**:710-717

[9] Meldman JA, Holt AW. Petri nets and legal systems. Jurimetrics: The Journal of Law, Science, and Technology. Dec 1971;**12**:65-75

[10] Specification and validation of protocols. IEEE Transactions on Communications. 1979;**1680**:1671

[11] Noe JD. A Petri net model of the CDC6400. In: Proc. ACMSIGOPS Workshop System Performance Evaluation. 1971. pp. 362-378

[12] Keller RM. Formal verification of parallel programs. Communications of the ACM. 1976;**19**:371-384

[13] Agerwala T. Putting Petri nets towork. IEEE Transactions on Computers. 1979:85-94

[14] Ramamoorthy CV. Performance evaluation of asynchronous concurrent systems using Petri nets. IEEE Transactions on Software Engineering. 1980;**SE-6**:440-449

[15] Cotronis JY, Lauer PE. Verification of Concurrent Systems of Processes, Tech. Rep. Univ. Newcastle-upon-Tyne; 1977. p. 97

[16] Baer JL, Gardarin G, Girault C, Roncairol G. The two-step commitment protocol: Modeling, specification and proofmethodology. In: Proceedings of the 7th International Conference on Software Engineering. 1980

[17] Zuberek WM. Timed Petri nets and preliminary performance evaluation. In: Proceedings of the IEEE 7th Annual Symposium on Computer Architecture. 1980. pp. 89-96

[18] Merlin PM, Farber DJ. Recoverability of communication protocols—Implications of a theoretical study. IEEE Transactions on Communications. 1976:1036-1043

[19] Shapiro SD. A stochastic Petri net with applications to modeling occupancy times for concurrent task systems. Networks. 1979;**9**:375-379

[20] Molloy M. On the integration of delay and throughput measures in distributed processing models [PhD dissertation]. Los Angeles: University of California; 1981

[21] Naktin S. Stochastic petri nets [thesis]. Paris, France: CNAM; 1980

[22] Amodeo L. Contribution to the simplification and control of stochastic Petri nets: Application to production systems [thesis]. France: Franche-Comté University; 1999

[23] Ascher H, Feingold H. Repairable Systems. Modeling, inference, misconceptions and their causes. New York: Marcel Dekker; 1984

[24] Stephens MA. EDF statistics for goodness of fit and some comparisons. Journal of the American Statistical Association. 1974;**69**:730-737

Chapter 7

ANFIS to Quantify Maintenance Cost of IT Services in Telecommunication Company

Leila Moradi and Reza Ehteshamrasi

Abstract

The maintenance cost predication of information technology (IT) regarding their important role and well-time availability in organization is valuable for IT managers. Therefore their decision originated from the predication might be great effect on organizational budgeting, planning, and strategy management. In this regard, having enough knowledge of IT system behavior and their cost forecasting may help IT managers to develop their organization. In this chapter, adaptive neuro-fuzzy inference system (ANFIS) with capability of modeling and predication is introduced, respectively, for quantifying information technology services and their maintenance cost in one of the telecommunication companies in Iran. Because of easy accessibility in finding parameters and also prevention from the complexity of information resulting from some available services in case study organization, automation services are selected by author as kind of user-involved and widely used in finding and studying on the variable data for implementation of the model.

Keywords: information technology, ANFIS, IT maintenance cost, IT services, IT cost forecasting

1. Introduction

Today, it is important to predict the cost of information technology in each organization for cost-effective purposes, and it is possible with the correct strategy and new modeling help technology managers to estimate the implementation costs and also intangible IT costs of the organization. One of the main issues related to the cost of information technology is the determination of costs, especially those that are of an indirect nature [1].

Terminating amount of investment and budgeting for each year, especially in new technologies including IT, is heavily complicated. IT creates a lot of cost for organization if IT managers do not have enough knowledge and information related to. Therefore, correct recognition of cost factors and affecting factors can be associated to cost prediction in the area and having great support to the cost-effectiveness of any enterprise. IT investment has a positive effect on technical effects on the firms' production process. The increasing dependence of many businesses on IT and the high percentage of IT investment in all invested capitals in business environment [2] are essential functions that should be considered within every organization, providing possible solution and services expected to assess the achievement of business objective. The first step in IT investment is to know exactly what that investment is and measuring and tracking this expenditure over time against a convenient base [3]. If the cost follows an industrial standard, it enables the organization to have right understanding for enough investment in the area of IT in a specific period. The growth of revenue via offered technology solution enables organization to achieve strategy business goals and level of competitiveness. In order to assess achievement level related to such expectation, mechanism has to be determined. Researchers and practitioners have expressed concern about cost, benefit, and quality of software documentation in practice. In this chapter, IT cost is estimated for system maintenance phase by quantifying IT practical services tangible by users. Having available information enables organization to determine how much maintenance cost has been spent for IT services with consideration of their sub-components. To achieve these goals, it is necessary to develop a model based on experimental and historical data utilizing ANFIS modeling. In the past decades, numerous studies have been published on software cost estimation method including expert judgment, parametric models, and at least machine learning [4]. The results of modeling are shown in which the ANFIS is used for quantifying generated consumable services of IT to show the level of effectiveness of IT activities and also their maintenance cost forecasting. Neuro-fuzzy inference system adapted to the Takagi-Sugeno known as a fuzzy model was used for these models. ANFIS is trained with a volume of data to quantify the services with fuzzy data. Four fundamental components utilized and applied in maintenance channel and system lost include human resource, technology/infrastructure, and process and system downtime considered as IT cost factors and normally tracked in any organization and feed to model. All four variables include some components with diversity and difference in each organization related to their activity.

This model can be suggested and designed as a new module implemented in monitoring systems of the use case organization considered as significant innovation with incorporation of IT services quantification and cost estimation, concurrently practical in IT field. The reason that authors used ANFIS is that it not only includes the characteristics of both methods but also eliminates some disadvantage of their lonely use case. In this chapter, backpropagation learning in neural section is applied. Output variables obtained by applying fuzzy rule to fuzzy set of input variables in Takagi-Sugeno inference system. The results showed that the proposed model is a powerful technique and valuable tool for forecasting variables from known and achieved knowledge that is not easily measured.

2. Theoretical framework

The prediction of maintenance costs is one of the important parameters in the budgeting of IT systems. It can help organizations with cost-effective management, including the intangible costs of technology and measuring the effectiveness and efficiency of technology services with regard to spending everyday expenses.

2.1 IT operating costs

2.1.1 Definitions and concepts

In information technology, operating costs refer to the amount spent on information technology services on a daily basis. Operating costs may be due to expenses that are spent on personnel, maintenance, hardware, power and software requirements, space rentals, and security.

Operating costs are often calculated annually or every 3 months. Some costs and benefits are simply measurable by using operational data and research techniques, which we call tangible costs, while some topics, such as intangible costs, may not be easily measurable and cannot be quantified in monetary terms [5]. The investment should be financially proportional to the function and profitability of the organization. Considering IT investment in organizations, costs, technical discussions, implementation tools, risk assessments, procurement strategies, and benefits should be taken into account.

2.1.2 Real costs of information technology

In information technology and systems, the actual cost of developing them is divided into two categories of direct and indirect costs [6].

2.1.2.1 Direct costs

The direct costs relate to the factors for implementation and operation of information technology; direct costs in projects are often underestimated and beyond hardware, software, and installation costs. Direct costs may also include additional hardware that is unpredictable. This includes increased processing power, increased memory and storage facilities, and the cost of installation and implementation, which are categorized as direct costs and typically include consulting services, installation engineers, and software and network hardware [7].

2.1.2.2 Indirect costs

Indirect cost includes two categories of human and organization. Indirect human costs include management time. Expenditures related to management time are specifically designed to integrate new systems into common activities. In addition, due to the creation of new technology, more time management should be spent on revision, approval, product reform, technology-related strategies, and intelligence systems. Organizational indirect costs include organizational issues that move from old-time activities based on the impact of new systems [7].

One of the main issues and difficulties associated with information technology costs is the recognition of costs, especially those that are of an indirect nature [1]. Indirect costs can be four times higher than direct costs. Some executive managers do not know clear understanding of indirect costs, and some seek to reduce the portfolio by attracting the manager support and at the same time increasing their portfolio value.

One of the biggest indirect costs is human time management. When new technology is created, administrators may spend time for revising, approving, and subsequently reforming IT-related strategies. Sample lifetime of backup costs is at least 400% of the initial purchase price. Then indirect costs arise from the development of new knowledge in the employee and the increase of flexibility and their overall contribution to the organization. Some indirect costs arise from employee payments and rewards that this staff turnover should be considered in the assessment of information technology [6]. Irani and Mohamed introduced a two-layer system for categorizing indirect human costs. The first layer is related to management, staffing, finance, and maintenance, and the second layer is introduced for specific cost elements. In the first layer of management, it decides how much of the expenditure is spent on information technology. Employees are those involved in information technology. Finance is the allocation of funds and maintenance cost for the development and implementation of technology.

2.1.3 Types of costs

Costs in information technology can be classified into two types of hard and soft costs. The hard costs are related to purchasing, maintaining, upgrading, licensing, and so on.

Software are a kind of cost that can be hardly measured, such as unplanned system downtime costs, very complicated design structures, having bad software and hardware, and so on. Soft costs are often based on hard cost.

2.1.4 Classification of costs

The classification of costs created is based on that fundamental infrastructure technology (this does not include outsourcing and the like). These costs are often uncertain and require special attention. Having a group of defined cost of information technology that includes enterprise information technology can provide ease of cost estimation.

2.1.4.1 The cost of manpower

The cost of human resources in information technology has to be taken into account in particular. Each record includes who is doing it. How much time does it cost and how much does it cost? The cost of human resources relates to those who hold the systems. Part of the cost is given to the software mentor as a backup contract to resolve the malfunction. We calculate the cost of training in this study for manpower costs. In the corresponding organization, about 3–4 courses are planned for training each year, each of course estimates 40–60 h of training. To promote the creation of a new system, training contracts with outside companies are also signed, which are not calculated by the organization.

2.1.4.2 Infrastructure costs

Cost is better to be hard cost as possible. It's very convenient for us to be involved with physical costs, including servers and locations, rather than labor costs. Labor costs that are a part of the cost of information technology can be addressed.

2.1.4.3 Facilities and buildings

This is the basis for the information technology infrastructure in a company. Companies often have one or more data centers. This feature of the cost includes building costs and installation costs for the implementation of information technology equipment. This classification is a useful indicator for comparing costs between two different sites. Stem elements include indoor items, including electricity, floor, and the like. Water can also be considered in this category because some data centers require water to cool their equipment. WAN connections include physical links from data centers to Internet providers and other communications required. Backup generators are often diesel cars that are outside of the datacenter.

2.1.4.4 Primary infrastructure

This level, with the resources available in the facility and building, provides an environment for information technology equipment. This includes a large portion of the cost to build a data center including SAN, NAS, network, and backup.

2.1.4.5 Computer hardware

The point to be made here is how to define a server to consider function and size of the server. Servers have different costs due to their different features and capabilities. Sometimes it's easy to set up virtual machines on a server for ease of use. Virtual machines work differently using a server's resources. Only some of the resources on these computers should be selected with the best and highest rating, including RAM, CPU, and so on.

2.1.4.6 Operating systems and backup software

2.1.4.6.1 Virtual software, operating systems, backup software

Often, operating system license fees can be specifically calculated, as server costs. And this also applies for virtual software. Backup software is software that is installed on operating systems and has license fees like antivirus.

2.1.4.7 Environmental infrastructure

This level of costs includes costs that are commonly used for many information technology infrastructures including software and hardware monitoring and management.

- Supervision
- Software and hardware management

2.2 Periodic cost of information technology

The periodic costs of technology are categorized according to direct and indirect costs by the following types.

1. **Training:** In some cases, the costs of training are known as direct costs but are estimated as indirect costs. For example, direct costs involve double training for a staff member by a vendor over many years, but indirect costs include new workforce training by existing personnel.

2. **Evaluation:** Direct costs include consultants that will assess the performance of existing systems as they are. Indirect costs include staffing responsibilities for aggregation and analysis of applied statistics.

3. **Upgrade and replace:** These costs are used for the future of information technology. Hardware and software applications should be upgraded, replaced, or licensed.

2.3 Maintenance and development in information technology

Maintenance involves supporting successful operating systems during long-term use by doing the following:

1. Regular and sample tests

2. Planned and preventive replacement of system units

3. Troubleshooting failure

4. Provided with spare parts

Maintenance activity is limited to determining the defective unit and its replacement or a spare part. Defective units are repaired at a central location equipped with tester a software control.

2.3.1 Maintenance activity

Lalli et al. [8] said that maintenance is divided into two parts, which include the replacement of defective elements. Moreover, preventive maintenance, which includes scheduled maintenance in the design phase, separates maintenance activity into three phases:

In Phase I: This includes planning and designing maintenance capability, which is part of a design process that involves the construction of elements that are easy to maintain and repair. In analyzing the measuring capability, the number of errors is often used from predictable times to maintain N's maintenance. Secondly, the expected time to maintain n is the expected maintenance with the program, as well as the expected time to repair and replace items that have a limited life span and a scheduled change program. Suppose the system is shut down during maintenance, then we can measure its availability.

In Phase II: This includes handling failure and replacing items that are short-lived as well as preventive maintenance operations.

In Phase III: This involves handling fault elements that specify whether it should be repaired or replaced.

2.3.2 Maintenance and availability relationship

The maintenance ability of system measures the effectiveness of a system in terms of ease and speed when the system and equipment return after failure. The possibility of a broken system can be restored to a definite period in operational conditions. Maintenance of availability becomes meaningful:

- If the system is interrupted and cannot be fixed in a minor way, we need to find the problem in the least amount of time.

- If spare parts are not available if they are needed.

- If repairs need more time to fix.

- If installing spare parts is very hard.

- If the testing and alignment of spare parts are difficult.

This makes the system inaccessible in operational areas (not operational).

Accessibility does not mean that only the existing elements should be in operation but that backup systems require that, although we have a definite unit. The availability of a system can be related to the reliability of the system and the ease of repairing the system and sufficient reservoirs for spare parts. For mathematical analysis, we assume that there is a system that requires continuous operation, except in cases where we need preventive and planned maintenance. In this case, we estimate the temporary backup system, or the system will be definite for a short time, but it may be operational at this time, even though some of its elements are not available. This kind of failure can be tolerated until further preventive maintenance.

The ultimate goal of a system's operation is that it is available when the system is needed. In systems, things such as failure rate, average repair time, one-piece delivery time, and operational constraints are all influence in accessibility. In the context of system maintenance analysis, review and measure the operating system and backup needs in quantitative and qualitative terms regarding design criteria in hardware and backup requirements [8].

In order to a system to be available, system providers should provide the required support over the life of the software and product. The ability to support the system requires an effective combination of reliability and maintenance and logistics and operational engineering.

2.4 Metric software measurement

Measuring the desirable performance and operational requirements of a computer system depends on our requests from the system. Performance measurement depends on three factors [9]

1. Slow activity

2. How does a system deal with unusual failure and conditions?

3. How does the system use the available resources?

When a computer system typically employs several types of activity, measuring the first feature is best done separately for any type of workload. Possible classifications for workload include timed checking, transaction processing, interactive computing, and categories. Errors create a variety of failures for the system, so all measurements in the category and the second floor must be evaluated for each type of error. For example, we should only use two types of error classes:

- Catastrophic

- Benign

Failure is visible effect of error. A catastrophic failure involves losing a significant amount of information, for example, a well-designed system. The CPU error may be benign, but the disk error may be catastrophic.

2.4.1 Measuring system performance

System performance is measured by the following assessment factors [9]:

1. **Responsiveness:** This measure evaluates how a task is executed rapidly by the system. Possible measurements include waiting time, processing time, conditional wait time (the wait time for an activity requires a certain amount of processing time), queue length, and so on

2. **Level of use:** This measure evaluates how a system uses its elements and components well. Possible measurements include operating power and the use of

different resources. The purpose of this measure is to measure responsiveness and differentiation and differences. Since a system that is well used is generally slower than the responsiveness system is less used.

3. **Mission capability:** This measure indicates that it is constantly operating in its mission period. Possible measurements may include the distribution of tasks performed during the mission, the availability distance (the system may function successfully during the mission), and the life span (when unacceptable behaviors increase beyond the boundaries of the mandate). This measurement is useful when repair and adjustment are unprofessional or acceptable disastrous behaviors.

4. **Reliability:** This measure determines how long a system will be trusted. Its possible measurement, including MTTF (average failure time) and MTTR (average repair time, long-term availability, and failure cost), is a useful measure when repair is possible and the failure is tolerable.

5. **Efficiency:** This measure determines how a user can effectively perform his work. Possible measurements include friendliness, maintainability, and user comprehension. Because measurement has a problem, it is often overlooked.

2.4.2 Availability

The data were presented by Culnan's three dimensions of accessibility as a framework [10]:

1. **Physical dimension:** Access to a terminal and the findings of an existing system (such as waiting time)

2. **Communication dimension:** the use of grammatical language to formulate a command including the ability to use the system and its command language and the ability to find a person to help for system use

3. **Information dimension:** The ability to recover physical information independently

Relative availability is the time when a system, or a component, should be operational at all when it is expected. A short phrase is that a system can continue to work even when the element or set of elements is gone. To measure reliability we must first measure availability.

Most reliability models are also predicted for hardware based on failure fracture rather than failures based on defects. In hardware, failure based on physical (corrosion) is more likely than design failure. The relationship between these key concepts of reliability in hardware and its application for software is specified. If we consider a computer system, a simple measurement of the reliability of the mean time between failures (MTBF) is displayed as follows.

In a transaction processing system, you may be concerned about system performance that is different from system levels, for example, consider the computer system one, A and the other is B. For the failure type, system A has an average time between two high failure MTTFs and has an average repair time (MTTR), but the failures are because of a large amount of files loss, and the system B has a short MTTF and a high MTTR. But loses of a small amount of data due to failure make user know that B is better than A although MTTF and MTTR and cost criteria may support the system.

One aspect of measuring the availability and reporting is definite scheduled time. To upgrade hardware, operating systems, databases and software, changing system and application parameters, data management, and backup to test failure and for any type of purpose, many programs require a permanent failure.

2.4.2.1 Measurement of availability

According to the theory, Stuart Rance and Hewlett Packard include the following steps:

1. Collecting data from the service desk that specifies the job impact and the duration of each event, which is often not expensive, but there is little discussion about the accuracy of the data.

2. The tool of all the elements needed to provide services and calculate accessibility is based on understanding how an element is combined with the back services. This is an effective case, but it may lose intangible failures. For example, a small database failure may be due to the inability of users to deliver specific types of transactions.

Use a group of clients that sends known transactions from a point in the network to determine what services are in progress and what they are doing.

According to Stuart Rance and Hewlett Packard, the concept of "user definitive minutes" often measures and reports the number of affected users that is based on multiplying the number of users by the time the system that is out of reach in minutes is being impressed. The number of "user definitive minutes" is often compared with the number of "user potential per minute" to calculate a number for accessibility.

$$\text{Availability} * \mathbf{100} = (\text{user failure minutes} - \text{user potential minutes}) / (\text{user potential minutes}) \tag{1}$$

Accessibility is an indispensable part of to be considered. It is very necessary to understand what is and how it affects and how it is calculated. Although many methods are proposed to calculate the availability of a device and a simple system, calculating the availability of a business application in a complex organization cannot be easily accessed [11].

2.4.2.2 Different types of accessibility

1. **Intrinsic accessibility:** When an item works satisfactorily in particular point under conditions in an ideal supportive environment, including logistics, waiting and regulatory failure, and preventive maintenance of the failure. This includes crashing maintenance. Intrinsic accessibility is generally due to the analysis of an engineering design, and the average of MTTF failure time calculation, which is divided by the average failure time plus the average time for MTTR repair based on design control, is calculated.

2. **Achieved accessibility:** When an item works satisfactorily in particular point of time under the declared conditions in an ideal supportive environment, for example, people, tools, and parts, which include logistic time, waiting, and

regulatory failure, which can lead to preventive and corrective maintenance of the failure.

3. **Operational availability:** When an item works satisfactory, is employed at a point in time and used in a reasonable logic operation and regulatory environment Including logistic time, preparation time, waiting, or regulatory failure.

It both contains corrective and preventive maintenance. Its calculation includes the average time between failures, which is divided by the sum of the average time between failure and the average MDT breakdown. This size defines the availability of elements that are controlled by logistics and extends planners, such as the quality of parts, tools, manpower, and hardware.

2.4.3 Usability of the system

Non-quality usability and the amount of information used to rely on resources [7, 12]. Culnan [10] has come to the idea that the requirements for collecting information related to the occupation, and occupation of each person may limit access to resources. The ability to use resources generally defines both the social and economic costs associated with communicating information. The online information system is evaluated in two dimensions, including usability and ease of use. Usability is related to the physical and informational dimension or grammatical language. When a new system is introduced, the user needs to be trained and supported to be able to get familiar with the system's command language. Usability can play an important role in influencing an individual to select an information source from among other suggested sources. Usability is likely to be affected by two types of field of use and initial experience [10].

2.4.4 Accessibility metrics and reliability

Reliability and availability metrics are used to measure them in a software product [9].

2.4.4.1 Average time to failure of (MTTF)

MTTF 500 means that a failure can happen every 500 times. The time units are completely system-dependent and can even be identified in a large amount of transactions. MTTF is related to a system with long transactions. When the system processes take a lot of time, MTTF should be longer than transactions.

2.4.4.2 MTTR (average time to repair)

When errors occur, the times are needed to fix the errors. Mean MTTR is the average time it takes to track and repair the errors that cause the system to fail.

MTTR is the average time to replace the elements that were defeated. When a hardware element fails, then the failure occurs permanently, and the MTTR will spend time repairing and replacing a new element.

2.4.4.3 Average time between two failures (MTBF)

To get MTBF, we need to combine MTTF and MTTR. A 300-h MTBF is a failure. The next failure occurs only after 300 h. In this case, time measurement is timely.

2.4.4.4 Possibility of failure in demand (POFOD)

A probability that a system fails against a request. In POFOD 0.001 it means that, 1 out of 1000 requests may lead to failure. The POFOD is an important measure for the security of critical systems and should be kept as low as possible.

2.4.4.5 Failure rate (ROCOF)

This metric is sometimes referred to as failure. The number of repetitions of behavior is unpredictable. ROCOF 100/2 means 2 failures that are likely to occur per 100 operating times. This is related to operational systems or process flow systems when the system should execute a significant amount of similar requests like a credit card process system.

2.4.4.6 Accessibility (AVAIL)

Availability makes it possible for a system to be available at a specific time. Availability of 0.998 means that every 1000 times the system unit is probably available for 998 times.

2.5 Performance evaluation

To evaluate performance, we have three techniques that are as follows:

1. Measure

2. Simulation

3. Analytical modeling

2.5.1 Measurement technique

An early technique is important for analysis, simulation, and modeling and works well for hardware, software, and linking patterns.

2.5.2 Simulation software technique

Simulation involves creating a model for the behavior of a system and advancing it with an appropriate look at the workload. The benefits of it are general and flexible, which should be considered in order to simulate the following:

It should be decided what the level of simulation does not simulate. Doubling the minor behaviors of a system is often unnecessary, and its high cost is forbidden.

Simulation as well as measurement also produces many raw data that should be analyzed using statistical techniques. Simulation is also necessary, as measurement by a carefully experienced design, to reduce the cost of maintenance.

In simulation, as well as measurement, only the behavior of a system is determined for the stench of inputs by the results obtained from data analysis.

2.5.3 Analytical modeling technique

An analytical modeling is the creation of a mathematical model of the system's behavior and its solution. Its advantages over the two previous ones are as follows: (1) creating a good attitude to the work of a system that is valuable, even if the model has a complex solution. (2) Simple analytic modeling can often be solved simply and obtains an exciting result. (3) The results of analytical modeling are much better than the previous two measurements, which can be predicted.

2.6 Adaptive neuro-fuzzy inference system (ANFIS)

The disadvantages of fuzzy inference systems and neural networks are the reason why the neuro-fuzzy systems appeared, retaining the advantages of both methods and outweighing the disadvantages. The lack of fuzzy inference systems is solved by creating the knowledge about a problem from the neural inference system training data, while the complicated and hard-to-understand rules of neural networks are bypassed by using linguistic variables by means of which results are easily explained. A well-known neuro-fuzzy system is the adaptive neuro-fuzzy inference system (ANFIS) used in solving various problems. The fuzzy inference system of Sugeno type can be considered as an adaptive neuro-fuzzy inference system in the form similar to neural networks in which by training the system on input/output data set, the parameters of the fuzzy inference membership functions or antecedent parameters and the parameters of the Sugeno fuzzy system output function or consequent parameters (pi, qi i ri) are adapted [13].

3. Methodology

According to the research type, data and research variables from the automation system of one of the state-run companies have been collected over several successive years for evaluating the services and the percentage of estimated costs in the maintenance area. These data are used first to determine the parameters of information technology services and then to the cost variables. This modeling sample is selected with experimental data as the representative of the total available data and allows us to generalize the simulation results to the whole model. The results of this research can be used to develop organizational monitoring systems. Data and information collected by using libraries, databases, the Internet, published article in conferences, scientific and research journals related to information technology systems, interviews with experts of the department of automation maintenance of the organization, observation factors influencing their measurement, the historical data collection of the organization, as well as the information were obtained by the experts error weekly for 1 year.

3.1 Introduction of measured variables

Accessibility is defined as the probability that a system works desirable at any random point of time. In order to make the system always available, various factors make the system out of accessibility, which can be measured the percentage of availability including the definitive system failure with planned reasons, and the other sudden system failure. The downtime of the system should be lowered, or the system will get out of order when it is needed [14–24].

3.1.1 System availability

Measuring system availability is a growing process that evaluates the behavior of computer systems and the increasing dependence of organizations on the frequent use of operating systems and the emphasis on the design of tolerance. In this study, the measurement of the availability of hardware and software and their effect on the failure of the service has been evaluated. In addition, four effective indicators of measuring this service have been evaluated for failure.

3.1.2 Data availability

The purpose of this variable is how data is available when stored in a particular way. It is often referred as storage resource through remote storage media. The speed of user's accessibility to the data and its access level in this variable are evaluated. The storage area network (SAN) is a network of external systems that communicates with the data source and network-attached storage (NAS) that stores data through the connection to the data network. Many factors are considered in this variable, including available bandwidth, security and availability level, file system type and access level, hardware and software storage type, and AST assigned to the system administrator. In this model, four effective indicators have been evaluated to measure the availability of data in the event of a malfunction and failure of the service.

3.1.3 Application response time availability

An operational request that can complete and perform the user's job requirements. This measure is used to analyze a general application of request and examines its operational numbers in relation to the ability to perform the required activity. Accessibility of request is part of software monitoring and management by observers to determine the system's ability to send requests for operational requirements. The measurement index in this research is based on the speed of the response of the application, the number of unsuccessful transactions and the responsiveness of the request, and other measurement metrics. Three input indicators has been evaluated to measure this service.

3.1.4 Accessibility to service and backup services

The availability level and quantitative results in each organization vary according to their type of activity. The organization may evaluate the availability of 99% for 87 h per year, but another organization has 99/99% availability for 53 h, in this research the findings evaluated regarding with interviews of the executive directors of organization based on the type of organization approved by the allowed fault per year. Regarding the availability level of backup services, one needs to know what makes an activity work when it's needed at a specific time. As defined by Gartner, the level of availability of backup services is related to "decreasing failure."

3.1.5 Network availability

The availability of network resources relates to the timely behavior of a network and network equipment, that is, when the system is uptime and can work satisfactorily at any point of time, especially when the system is in repair. Accessibility is the key feature to any organization that exchanges vast amounts of data in their databases, and it is important in critical measurements, but it is important for daily operations. Some failures sometimes rise to concerns for network administrators. It may require some changes, but today, in large organizations, backup equipment is needed to reduce the risk of loss. In the studied organization of this research, the existence of backup equipment has reduced the downtime and raised availability of the system in this regard. There are four numbers of indicators measured in this study for this service.

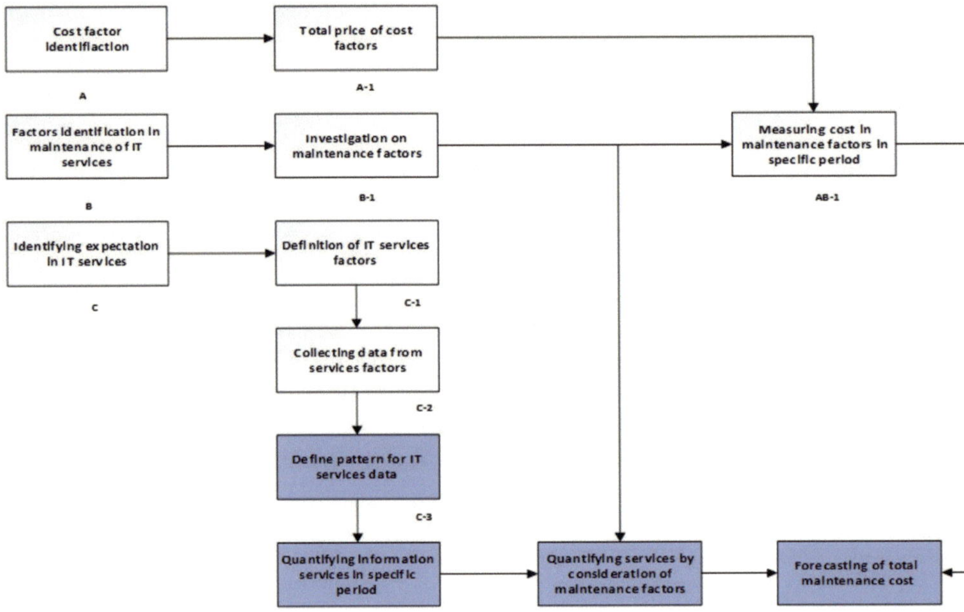

Figure 1.
Workflow diagram.

3.1.6 Turnaround time availability

The amount of time spent to do a specific request. The turnaround time is total time between sending a program, a processor, to a complete output for the client and the user. And sometimes it may return to the total time required to complete a process. And it can provide the expected output to the user after the program execution. An indicator turnaround time is used to evaluate the operating system programming algorithm. Here is the period of time a request is executed from the user's side.

3.2 Method

Adaptive neuro-fuzzy inference system (ANFIS) has been used to determine the quantity of automation ACS services by finding their indicators in the organization. Indicator data are derived from the available data in an effective organization, which finds the parameters of these indicators in measuring the level of service availability. And in each organization, there are different indicators based on the type of service being studied (**Figure 1**).

4. Research modeling and finding

4.1 Identify effective factors on the availability of IT services in office automation

The automation systems of the organization are evaluated based on availability for the user at normal and critical times, due to the service failure time. In calculating the availability metric for the above service, six main variables are defined as output for this service. Each of these outputs has different factors to check availability. Because these outputs were not measurable in organizations alone, their factors that contributed to its measurement searched for and evaluated for each of them.

Availability of consumed services in IT	Symbol	Number of attributes	Index measuring
System availability services	ACS1	4	min
Data availability services	ACS2	4	min
Application response time availability services	ACS3	3	Sec/millisec
Service support availability services	ACS4	3	min
Network availability services	ACS5	4	Min/milllisec
Turnaround time availability services	ACS6	3	min

Table 1.
Component of availability of IT services.

4.2 Find elements affecting the output

We discussed six main and effective variables on the measurement of the availability of information technology services (**Table 1**). ANFIS modeling with six below major variables with the abbreviations marked in front of them was written in the research. It consists of six elements:

These six major variables of research on information technology services were found and analyzed on automation services. Each of the six main variables has several sub-elements or indicators in collecting data and analyzing them. These indicators are events that assess the level of service availability from the point of view of definite time or out of service. The events from the point of view of failure and the number of endings in weekly periods are examined to predict the variable availability according to the time expected for the operation of a system (AST).

4.3 ANFIS to estimate maintenance costs for office automation

In the second modeling, ANFIS is used to predict the costs of maintenance of office automation service over a specific time by consideration of the fixed maintenance costs and the costs of the failure of the service, and the parameters obtained from the accessibility of the services in the first model weekly for these services were taught by the model. Given the importance of some indicators in measuring cost of service failure, the amount of these factors was determined and included in the cost factor.

$$\begin{aligned} Cost = \{&First\ variable\ failure\ cost + Second\ variable\ failure\ cost \\ &+ Third\ variable\ failure\ cost + Fourth\ variable\ failure\ cost \\ &+ Fifth\ variable\ failure\ cost + Sixth\ failure\ cost\} \\ &+ \{Cost\ of\ fixed\ maintenance\} \end{aligned}$$

$$(cost) = f(lost\ time\ (ACS1 + ACS2 + ACS3 + ACS4 + ACS5 + ACS6)) + fix\ cost$$

Although the maintenance cost represents a very small percentage of the company's revenue, the software and service failure rate, due to the amount of defects

Test number	Real cost (ml/r)	Forecasting cost (ml/r)
1	436	424
2	732	746
3	685	689
4	667	662

Table 2.
Output table of the model in four times of testing.

at intervals, and the decrease and drop in speed, have a great effect on increasing the cost of maintenance in the organization studied leading to more work hours and replacing faulty components and increasing the time of support contracts to minimize the criticality in the service data shown in (**Table 2**).

4.4 Summary

The availability of services was determined by the quantitative characteristics of measurable indicators in the organization as inputs to the model. These characteristics were selected after the analysis of the results of the questionnaire with the senior organization's experts and data recorded in the monitoring system of the maintenance unit. The error obtained from actual data comparison with the predicted model data was confirmed by managers and experts in the organization.

The variables of measurement indicators in system maintenance cost include human resources, equipment training, and processes and failure rates, in which behavior of each indicator in an organization varies depending on its type of activity.

5. Conclusion

Nowadays, the measurement of accessibility and availability of IT services for IT administrators and also consideration of their associate costs are essential factors for future-oriented decisions from the capability of the system based on user expectation to accurate investment in the field. Accurate investment in IT is complicated for some IT administrators based on the number of current services in their organization. Identification of new artificial intelligence methods will help them to know the level of effectiveness of IT services in their organization in each period. In the present study, to prove and verify proposed prediction framework, MATLAB fuzzy logic toolbox is used.

This tool provides us with ANFIS as a selected learning technique to present and develop the model in two phases: first phase is quantifying the level of IT services' availability (six variables), and second phase is prediction of their maintenance cost of system failure by finding of six variables' measurement of services. To prevent any complexity for data collection resulting from large number of services, automation system placed in system maintenance department of study organization was selected as a user-involved and widely used system to find out indicators, affective reasons, and implementation of the model. And also this department helps to estimate lost maintenance costs and find its effective indicators. Regarding the result of this study, ANFIS can predict well with lowest fault and near to real data. It is kind of effective, new, practical technique than others with precise prediction.

Table1: Inputs and outputs of ACS1

MF	HT	HDF	NDF	Target	Output
87.3	100	95.2	89	93.1	94

Table 2: Inputs and output of ACS2

TS	CL	LCS	NR	Target	Output
99.6	97.3	96.5	92.2	97.3	97

Table 3: Inputs and output of ACS3

EM	GT	HR	Target	Output
99.9	99.9	99.7	99.9	99.9

Table 4: Inputs and output of ACS4

RTP	RTN	RTO	Target	Output
62.5	89.8	100	98.3	97.9

Table 5: Inputs and output of ACS5

PT	ST	ND	BT	Target	Output
98.2	96.5	95.7	99	96.2	96.3

Table 6: Inputs and output of ACS6

TI	AR	DR	Target	Output
85.5	86.7	72.2	83.7	84.7

Table 7: Inputs and output of cost

ACS1	ACS2	ACS3	ACS4	ACS5	ACS6	Target	Output
93.1	97.3	99.9	98.3	96.2	83.7	1411	1419

Table 3.
Output of ANFIS training.

As for future work, (1) the model can be developed as a basic suggestive model to measure other kinds of cost-related service quantification; (2) it is practical for the understudy organization to develop a management module in monitoring systems by combination of both models' targets discussed in the article. It provides managers with views to check constantly the status of the system and the reasons of increased cost to be applied in future decisions; (3) more advanced method can be involved in the experiment as regression; and (4) extensive experiment on more variables and attributes can contribute to more realization.

Table 3 shows the outputs of the model with respect to forecasting of cost and quantifying IT services studied in automation services in the organization in specific periods.

Author details

Leila Moradi[1]* and Reza Ehteshamrasi[2]

1 University of Science and Research (Islamic Azad), Tehran, Iran

2 Department of Industrial Management, Qazvin Branch, Islamic Azad University, Qazvin, Iran

*Address all correspondence to: Moradi.lyla@gmail.com

References

[1] Peter ED. Love, Ahmad Ghoneim Zahir Irani. The Journal of Enterprise Information Management. 2004;**17**(4):312

[2] Jafari SM. Strategic cost-cutting in information technology: Toward a framework for enhancing the business value of IT. Iranian Journal of Management Studies (IJMS). 2014;7(1):21-39

[3] Dolci PC, Macada ACG, Grant G. IT investment management and information technology portfolio management (ITPM). Journal of Enterprise Information Management. 2014;**27**(6):802-816

[4] Huang J, Li YF, Xie M. An empirical analysis of data preprocessing for machine learning-based software cost estimation. Information and Software Technology. 2015:108-127

[5] Osiro YW. A model for costing information technology services in public organization (thesis). Strathmore University; 2012

[6] Love PED, Ghoneim A, Irani Z. Information technology evaluation: Classifying indirect costs using the structured case method. Enterprise Information Management. 2004:312-325

[7] Irani Z, Ezingeard JN, Grieve RJ. Costing the true costs of IT/IS investment in manufacturing: A focus during management decision making. Logistics Information Management. 1998:38-43

[8] Lalli VR et al. Designing for Maintainability and System Availability. National Aeronautics and Space Administration; 1997

[9] McGraw-Hill. Introduction to Computer System Performance. 3rd ed. Singapore: Library of Congress Cataloging—in Publication Data; 1992

[10] Culnan MJ. The dimension of perceived accessibility to information: Implications for the delivery of information system and services. American Society for Information Science. 1985:302-308

[11] Rohani H et al. Calculationg Total System Availibility. France Amsterdam: Information System Organization KLM-Air

[12] Sonmez R. Range estimation of construction costs using neural networks with bootstrap prediction intervals. Expert Systems with Applications. 2011:9913-9917

[13] Svalina I, Galzina V, Lujic R, Simunovic G. An adaptive network-based fuzzy inference system (ANFIS) for the forecasting: The case of close price indices. In: R.285. Expert System with Application. Elsevier; 2013. p. 9

[14] Dugan RE. Information technology budgets and costs: Do you know what your information technology cost each year? The Journal of Academic Librarianship. 2002:238-243

[15] Majazi Dalfard V, Nazari Asli M, Asadzadeh SM, Sajjadi SM, Nazari-Shirkouhi A. A mathematical modeling for incorporating energy price hikes into total natural gas consumption forecasting. Applied Mathematical Modelling. 2013;**37**:5664-5679

[16] Mavaahebi M, Nagasaka K. Quantifying information technology's generated services and incurred costs by applying empirical artificial neural network/Expert System Modeling. In: International Conference on Advanced Mechatronic Systems; Luoyang, China; 2013. pp. 25-27

[17] Marsh L, Flanagan R. Measuring the costs and benefits of information technology in construction. Engineering

Construction and Architectural Management. 2000:423-435

[18] Marandi AK, Khan DA. An impact of linear regression model for improving the software quality with estimated cost. Procedia Computer Science. 2015;**54**:335-342

[19] Prasad K, Gorai AK, Goyal P. Development of ANFIS model for air quality forecasting and input optimization for reducing the computational cost and time. Atmospheric Environment Journal. 2016:246-262

[20] Pitt LF, Watson RT, Kavan CB. Measuring information systems services quality concern for a complete canvas. MIS Quarterly Academic Journal. 1997;**21**:P209-P221

[21] Bosse S, Splieth M, Turowski K. Multi-objective optimization of IT service availability & cost. Reliability Engineering and System Safety. 2015

[22] Ezran P, Haddad Y, Debbah M. Availability optimization in ring-based network topology. Computer Networks. 2017:27-32

[23] Arnaiz A, Konde E, Alarcon J. Continuous improvement on information and on-line maintenance technologies for increased cost-effectiveness. In: 2nd International Through-Life Engineering Services Conference Journal; Elsevier; 2013. pp. 193-198

[24] Daneshkhah A, Bedford T. Probabilistic sensitivity analysis of system availability using Gaussian processes. Reliability Engineering and System Safety. 2013:82-93

Chapter 8

On-Field Operation and Maintenance of Photovoltaic Systems in Cameroon

Kodji Deli and Djongyang Noel

Abstract

The objectives of this work are to examine the causes of the breakdown in the photovoltaic power systems, to propose strategies to solve them, and to evaluate the field lifetime of some elements of the PV systems. The data analyzed were obtained from maintenance records and measurements over a period of 9 years (from 2007 to 2015) for the backup PV systems and 2 years (from 2016 to 2018) for photovoltaic water pumping (PVWPS) systems. It appears from this analysis that 29% of the batteries went bad (leading to curative intervention); this contributed to about 64.9% of the total breakdown registered. Using the failure modes, effects and criticality analysis (FMECA) method for PVWPS, criticality is 252, 402, and 504 for inverters, PV module, and motor pump, respectively. This demonstrates that motor pumps are more sensitive than other elements in the PVWPS. This study also permitted not only to evaluate the quantity of preventive and corrective maintenance impacts on solar PV systems but also to propose maintenance strategies to rapid diagnosis of PV systems.

Keywords: breakdown diagram, life expectancy, maintenance strategies, backup PV systems, PV pumping systems

1. Introduction

The production of energy is an immense challenge for the coming years. Indeed, the energy requirements for industrialized societies are increasing [1]. Nowadays, 80.9% of world production of primary energy is supplied from fossil resources [2]. The scarcity of conventional energy and the environmental problems caused by its use have led to the usage of other renewable sources such as solar photovoltaic energy. It is stimulated first by the availability of solar resources in most part of the globe particularly in Africa where there is a strong solar resource and secondly by the decrease in the cost of photovoltaic equipment during the last decade [3], an average of 0.7 $/kWh in 2016 and 0.5 $/kWh in 2020. The production capacity of solar photovoltaic energy within the last three decades has witnessed a yearly increase of 44.2% between 1990 and 2010, to reach a production capacity of 99.2 GW in 2012 [4]. Global installed PV capacity at the end of 2016 was reported as 310 GWp [5]. The price of photovoltaic module dropped by 80% between 2009 and 2015 to reach the actual cost which is less than 1 USD/Wp [6]. PV is widely used in many applications nowadays [7].

The use of renewable energies has increased significantly in Cameroon these recent years since it is demonstrated that the access to modern forms of energy can contribute effectively to the revival of economy and reduction of poverty. In many countries around the world, the use of renewable energy contributes expanding employment opportunities which lead to promoting human development [8]. Recently Cameroon has embarked on the use of renewable energy, which has led to the creation of a directorate of renewable energy in the ministry of energy and water. Investments have been made in the public investment budget (PIB) for the installation of renewable energy systems particularly solar energy [1], for example, public lighting in cities and the countryside by using solar street lights, solar power plants for the villages' electricity supply, battery charging stations in villages, and solar power supplies for community centers. However, as they are installed in outdoor environment, continuous exposure to harsh environmental conditions (sun beam, rainfall, etc.) may reduce the optimal performance of the system. PV systems are difficult to implement because they encounter problems among which is the problem of servicing and maintenance. An effective operation and maintenance (O and M) program enables PV system production to reach its expected level of efficiency, which will consequently strengthen end users' confidence in such systems [9].

The field performance of photovoltaic systems has been extensively studied for many applications especially in countries with strong database of solar resource [9]. However, these databases are used exclusively for assessing the electrical performance of the system [10]. To model the annual performance of photovoltaic modules, their performance characteristics are needed [11, 12]. The available information from manufacturers are typically limited to temperature coefficients, short circuit current I_{sc}, open circuit voltage V_{oc}, and maximum power P_{max}, at rating conditions (G = 1000 W/m^2, Tc = 25°C, AM = 1.5). The information is useful when one want to compare photovoltaic module performance at rating conditions but are inadequate to predict annual field performance under typical operating conditions [13]. It is demonstrated that there is difference between expected power production forecasts and field experience of photovoltaic arrays [14]. It has been shown that the relative performance ranking at rating conditions may not agree with the ranking based on monthly or annual performance. Faults in PVS may cause a huge amount of energy loss. A monitoring study was conducted on a test PV system by Firth et al. [15], and it was reported that the annual power loss due to various faults is about 18.9%.

Failures that occur in the PV systems can cause system shutdown. The main components involved are PV modules, cabling, protections, converters, and inverters. Failures are mainly caused by external operating conditions which are shading effects, module soiling, inverter failure, and aging of PV modules [16]. The line-to-line fault (LLF), ground fault (GF), and arc fault (AF) are tree catastrophic failures encountered in PV arrays [17]. PV system maintenance and performance are related to good inspection and monitoring. These are important in determining life-cycle costs and servicing requirements. Photovoltaic energy is seen as a viable option for decentralized energy production; the sustainability of these systems does not only depend on the initial system cost but also on the cost of maintenance and the lifetime related to the maintenance operations used [18, 19]. This chapter presents an overall of existing faults encountered in both DC and AC sides over a period of 9 years in more than 20 PV systems in Cameroon; this chapter also proposes detection techniques with a fault detection procedure (the breakdown tree diagram) that is intended to facilitate interventions on all components of PV systems.

2. Maintenance strategies

Maintenance strategies are the "heart" of the maintenance planning process. They are responsible for defining the "maintenance actions" based on the information obtained from the system and preprocessed. Maintenance strategies are corrective, preventive, condition-based, opportunistic, focused-on-reliability, and production strategies [20]. The questions when, what, who, where, why, and how are the system interventions that should be executed or not, in order to keep the system functions alive [21]. One of the maintenance objectives is to reduce the failure occurrence, increase the availability, and extend the system life (or at least in the mean time until the next failure).

3. Maintenance technique management

A maintenance schedule, planning, and management are important for the evaluation of the health condition components and the incipient fault diagnosis. Different aspects of the operation and maintenance of renewable energy systems were proposed by [20, 22]. A generic structure of asset management which integrates business decisions to optimize investment decisions related to maintenance is presented by [23], and it consists of eight blocks of sequential management. In physical asset management, the maintenance optimization is a concern, because in general, the assets deteriorate as it is being raised and both the failure risk and cost increase [24]. Maintenance management model of assets is presented in **Figure 1** below.

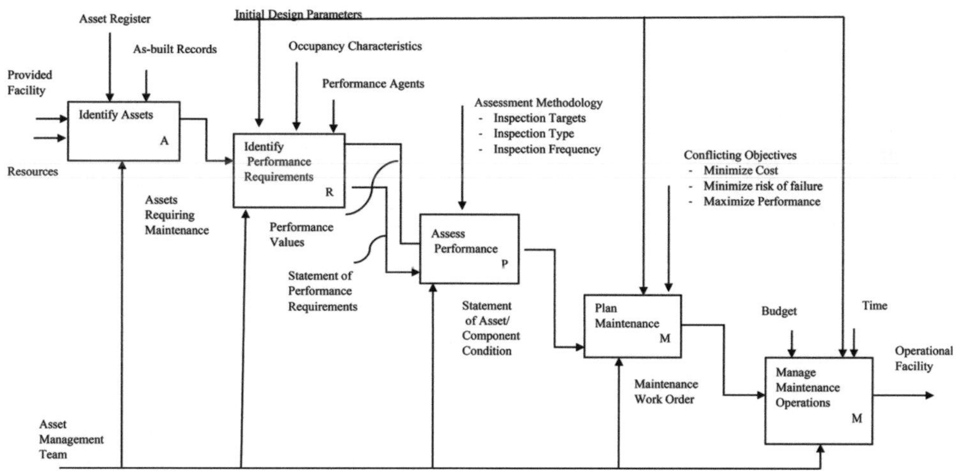

Figure 1.
Maintenance management model [25].

4. Operation and maintenance of the photovoltaic systems

4.1 Preventive maintenance

Preventive maintenance consists of a regular observing passage and a frequent replacement of exhausted constituents of the system. Preventive maintenance can be systematic or conditional. Systematic preventive maintenance consists of changing worn out materials according to a preestablished schedule [26]. Preventive maintenance is scheduled at regular time intervals, independent of component wear, or if it still executes its function satisfactorily [26]. Preventive maintenance scheduling could be done using several strategies; the most common are usually with a minimal cost target that are based on the budget allocation for maintenance in accordance with the system priorities [27, 28]. The objective of this type of maintenance is to ensure the reliability of the system and to maintain the system in its state of initial efficiency [29].

4.2 Corrective maintenance

It consists of the setup of a breakdown system. It is usually done in two stages: palliative corrective maintenance (fixing) which involves the start off of a system which is partially or totally broken up while waiting for a permanent repair of this system. In this case, the speed with which interventions are done is considered, and the action must take place as fast as possible for quicker start off of this system [20]. The goal of this action is not to repair the breakdown but to permit the system to fulfill part or the totality of it function [20]. Curative corrective maintenance (repairing) is a final setup of all the worn out elements of the system. Contrary to the fixing action, the repairing action is a planned one. In this case, the quality of the intervention is more important than speed. In the case of corrective maintenance on photovoltaic systems, diagnosis diagrams are first done in order to ease and help workers to determine the worn out elements knowing the causes of the breakdown. The essence of the approach "run-to-failure" or corrective maintenance is to replace the component with a new one when it is not able to perform its function [26].

4.3 PV systems fault detection techniques

There are several techniques for the detection of faults in PV systems; these techniques are summarized in **Figure 2**. These techniques have helped in improving the system reliability and lifetime of PV systems. The classification of different fault

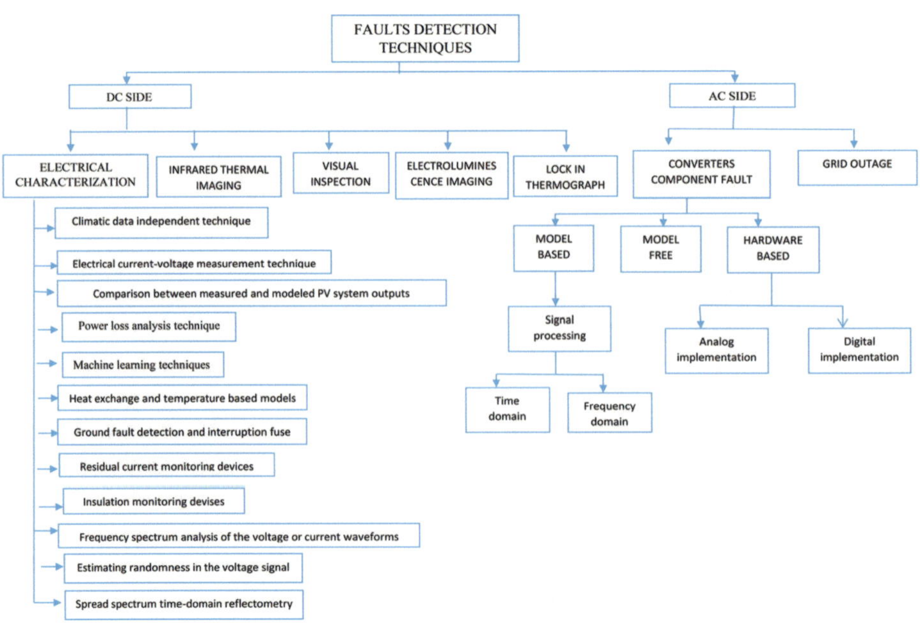

Figure 2.
Fault detection techniques in DC and AC side of PV system [27, 29].

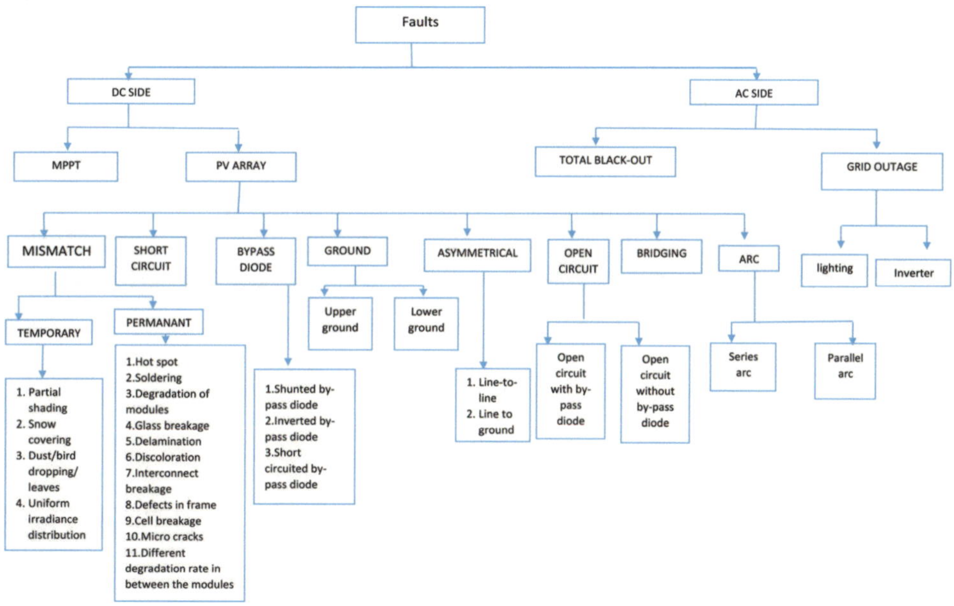

Figure 3.
Classification of faults in DC and AC side of PV systems [27].

detection techniques to identify the type and location of the fault occurring in DC and AC sides of PV system is given in [27, 16, 28]. Although other techniques are simple to implement, most of them require monitoring and analysis of the electrical performance of photovoltaic (PV) systems.

4.4 Types of faults encountered in PV systems

In the field conditions, a number of factors can cause a PV array to reduce its output power. Any factor which reduces the output is considered as "fault" [27, 30]. Generally, faults in PV systems can be classified into two main categories: permanent and temporary. The classification of the most common types of fault in PV system is presented in **Figure 3**. The main faults encountered in PV systems Installed are presented in the **Figure 4** for shading, **Figure 5** for soiling and dust, **Figure 6** for PV jonction fault and discoloration, **Figure 7** for wiring fault, **Figure 8** for circuit breaker and inverter, **Figure 9** for sulfatation and deep discharge of batteries.

5. Characteristics of the studied PV system

5.1 Backup PV systems

To overcome the problem of power cuts, 20 backup systems have been installed in 20 cities in Cameroon at the end of 2006. These systems are essentially made up of monocrystalline solar modules (Hélios, 80Wc), batteries (Midac, 400 Ah) to backup during power cuts, and charge controllers (Steca, 20A) which regulates the energy flux and protects the batteries from overloading and deep discharge. There are generally three types of controllers [31, 32]: shunt controllers, series controllers, and maximum power point tracking (MPPT) system. The inverter called "inverter-chargers" can be connected to the

(a) (b)

Figure 4.
(a) Shading and (b) vegetation and building.

(a) (b)

Figure 5.
(a) Soiling and dust accumulation and (b) vegetation.

(a) (b)

Figure 6.
(a) PV junction box and (b) delamination and discoloration.

electricity network in the purpose of supplying the energy of the network, two types of inverters are used C1600-12 and C2600-24 (Studer compact) (**Figures 4–9**).

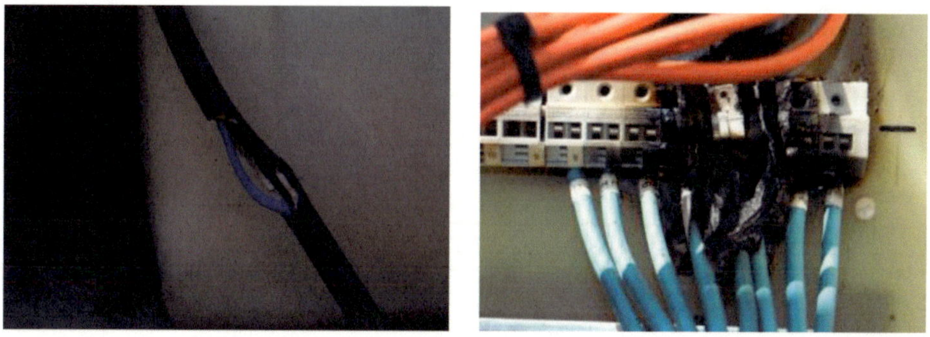

(a) (b)

Figure 7.
(a) Wiring system fault and (b) poor tightening of connections fault.

(a) **(b)**

Figure 8.
(a) Circuit breaker fault and (b) inverter fault.

(a) (b)

Figure 9.
(a) Deep discharge of batteries and (b) sulfation of batteries due to lack of maintenance.

5.2 Photovoltaic water pumping system

Two water pumping systems are installed to provide water to the population of an isolated site located at 10°23 N and 14°26 E. Each system has 13 kWp of PV generator associated to an automatic inverter. The pumping takes place over the

Denomination	Symbol	Value
Rated maximum power (P_{max})	P_{max}	255 Wp
Maximum power current (Imp)	I_{max}	8.13 A
Short circuit current (I_{sc})	I_{sc}	8.61 A
Maximum power voltage (V_{mp})	V_{mp}	31.52 V
Open circuit voltage (V_{oc})	V_{oc}	37.92
Maximum system voltage		1000 V
Operating temperature		(−40 + 85)°C
Temperature coefficient of I_{sc}	α_{Isc}	−0.58%/°C
Temperature coefficient of V_{oc}	β_{Voc}	−0.33%/°C
Temperature coefficient of power		−0.41%/°C

Table 1.
Characteristics of the PV module used in PVWPS.

Inverter Solartech PB11KH		Grundfos SP17-10 pump	
Denomination	**Value**	**Denomination**	**Value**
Rated power	11 kW	Motor type	MS4000
Max solar input power	16 kW	Rated power—P2	5.5 kW
Input string	4	Power (P2) required by pump	5.5 kW
Max input current of each string	15 A	Mains frequency	50 Hz
Max DC input voltage	750 V	Rated voltage	3 × 380–400–415 V
Recommended MPP voltage	500–600 V	Rated current	13.0–13.0–13.4 A
Adapting motor power	9.2–11 kW	Starting current	480–530–550%
Adapting motor voltage	3PH 380-440 V	Cos phi—power factor	0.85–0.81–0.76
Rated AC output current	24A	Rated speed	2850–2860–2870 rpm
Output frequency	0–50/60 Hz	Starting method	Direct online
Conversion efficiency	Max 98%	Rated flow	71 m^3/h
Ambient temperature	−10 to 50°C	Rated head	81 m

Table 2.
Characteristics of the inverter and the motor pump.

sun. The configuration of a field is performed as follows: 51 solar panels all divided into 03 strings of 17 panels. Two pumps P1 and P2 were associated. The characteristics of the PV module used, inverter and motor pump, are specified in **Tables 1** and **2**.

6. Maintenance and fault detection techniques in PV systems

The field inspection process is a key to the development of healthy and safe PV systems. Many works consolidated the most important aspects of a field inspection of photovoltaic system which is the competency of the contractor and installer without taken into account life service of each element and their implication on the failure of the system [33, 34]. Diagnosis procedures consist on visual inspection procedures (array inspection, wire inspection, inverter inspection, inspection of module, and array grounding) and performance monitoring (performance verification, displays, design software, data acquisition systems, sensors) [34]. One of the most valuable techniques for identifying existing problems and preventing future problems is to walk to the site and conduct a thorough visual and hands-on

Parameter	Symbol	Unit
Meteorology		
Total irradiance (global), in the plane of the array	G_I	W m^{-2}
Ambient air temperature in a radiation shield	T_{am}	°C
Wind speed (may be required by special contract or if the PV array is subject to extreme operating conditions)	S_W	m s^{-1}
Photovoltaic array		
Output voltage	V_A	V
Output current	I_A	A
Output power	P_A	kW
Module temperature	T_m	°C
Tracker tilt angle (optional for systems with tracking arrays)	ϕ_T	Degrees
Tracker azimuth angle (optional for systems with tracking arrays)	ϕ_A	Degrees
Energy storage		
Operating voltage	V_S	V
Current to storage	I_{TS}	A
Current from storage	I_{FS}	A
Power to storage	P_{TS}	kW
Power from storage	P_{FS}	kW
Load		
Load voltage	V_L	V
Load current	I_L	A
Load power	P_L	kW
Utility grid		
Utility voltage	V_U	V
Current to utility grid	I_{TU}	A
Current from utility grid	I_{FU}	A
Power to utility grid	P_{TU}	kW
Power from utility grid	P_{FU}	kW
Backup sources		
Output voltage	V_{BU}	V
Output current	I_{BU}	A
Output power	P_{BU}	kW

Table 3.
Parameters to be measured in PV systems in real time [35].

inspection of the PV system components. During these inspections, the parameters to be measured in real times are specified in **Table 3**.

When problems are identified, we can use breakdown tree diagrams [36, 37]. Breakdown tree diagram gives a graphical description of the different events that lead to a breakdown resulting to the non-reliability and the stop of the system [37]. The breakdown tree diagram is constructed in a deductive manner. It starts with the peak event right up to the elementary event in arborescence. The peak event for which we seek the probability is often called "feared." We generally use AND and OR logic gates to define the probability of what is at the cause of the event, to put the situation (what is to be resolved) at the head of the diagram and link it to its causes (events that can be at the origin) by the gate. Once the diagram has been archived, if there is any breakdown, interventions are done from the bottom of the diagram to the top where the problem is detected in the system. **Figure 10** shows that for the "feared" event to archived, either event E1 or E2 must have been archived. In the same manner, for the event E1 to be archived, either the base event e1 or e2 must have been archived, and for the event E2 to be archived, both base events e1 and e2 must have been archived at the same time.

The failure modes, effects and criticality analysis (FMECA) which is a rigorous and preventive method for identifying potential failures of a system and elements, actions have been defined to be taken to eliminate these failures, reduce their effects, and detect and prevent causes. The method is part of an eight-step process [38] as seen in **Figure 11**. Several criteria can be used to determine the criticality index. In practice, we assign three notes (each on a scale of 1–10) for each trio cause-mode-effect:

- The grade G: severity of the effect, the consequences on the client/user
- The grade O: the probability of occurrence, the frequency of occurrence
- The grade D: the probability of non-detection, the risk of non-detection

$$\text{The criticality index is obtained by } C = G * O * D \tag{1}$$

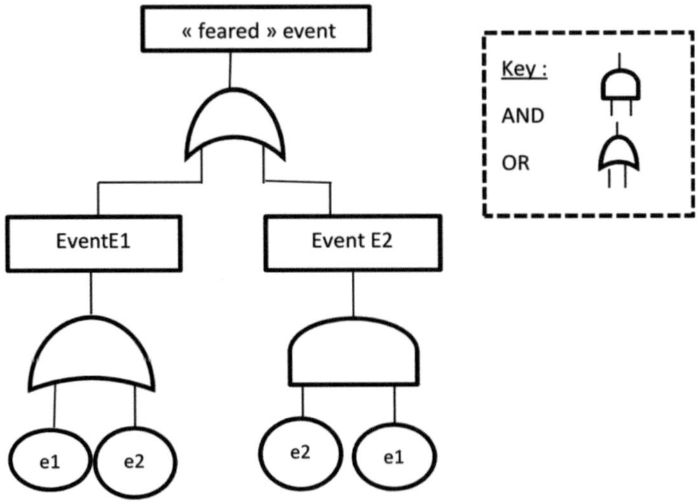

Figure 10.
Breakdown tree method.

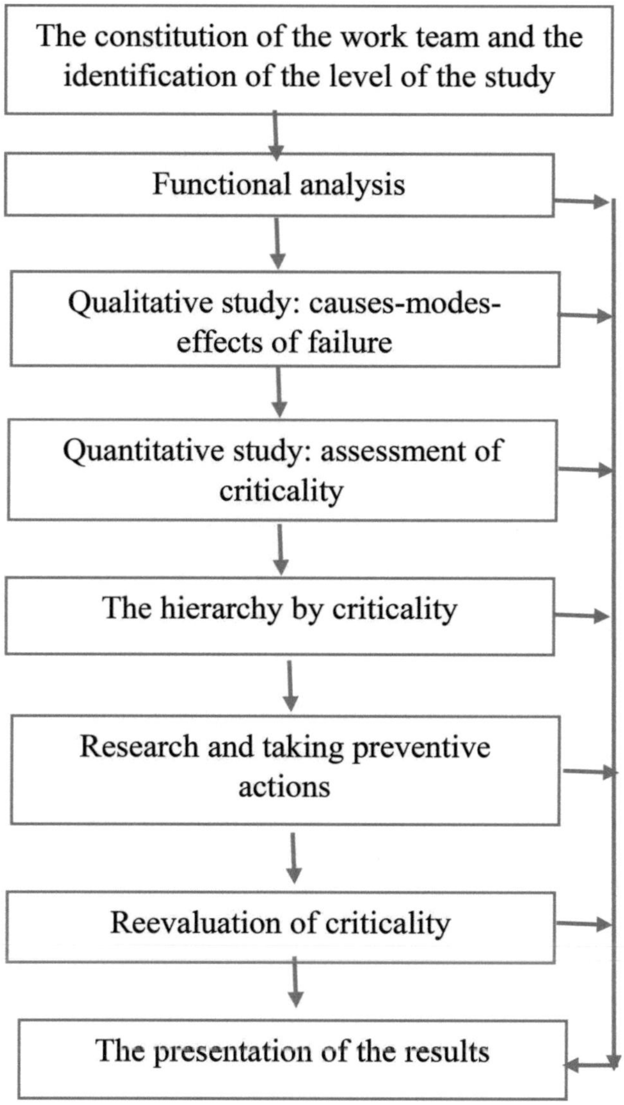

Figure 11.
FMECA approach.

7. Case study

7.1 Photovoltaic backup systems

In order to better study the impact of the two types of maintenance on the system, **Figure 12** shows the frequent preventive and corrective maintenance operations carried out on installed PV backup systems during the period 2012–2015, which were kindly followed.

Figure 12(a) presents the number of preventive and corrective interventions realized on one site within a period of 4 years. This illustration shows the importance of preventive maintenance on PV systems. In effect, the more preventive maintenance are done, the less there are corrective operation realized. It is the case for the years 1, 3, and 4. To estimate the element lifetime before failure, the exploitation of the maintenance files indicated that 25 batteries were damaged on the 82 installed as shown in **Figure 12(b)**; thus, batteries contributed to 64.9% of the breakdowns

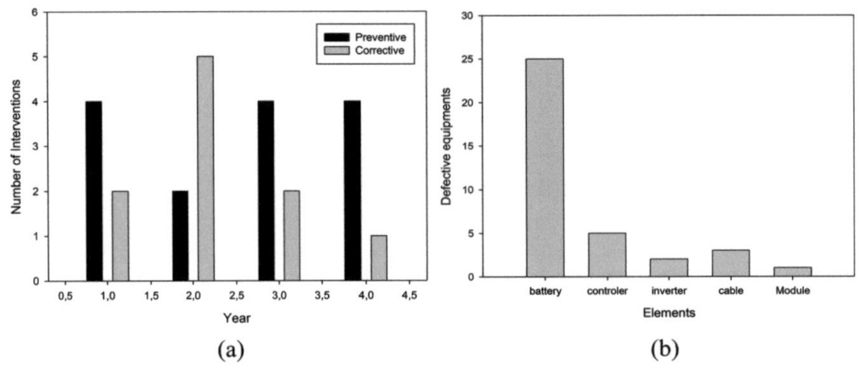

Figure 12.
(a) Number of corrective and preventive maintenance per year on a PV systems and (b) recorded breakdowns per elements.

generated during this studying period. This result shows also that batteries are the most sensitive elements of the PV system charge controllers, cables, and inverters, which contribute, respectively, 13, 8.3, and 5.5% of the breakdown registered.

7.1.1 The causes of breakdown at each element

7.1.1.1 Causes of the battery and cable breakdown

Batteries being one of the vulnerable elements of a PV system, the direct causes of their breakdown are due to late interventions and aging. **Figure 13(a)** shows the causes of the breakdown of batteries. It appears that five battery breakdowns were caused by the lack of control of the good functioning of charge controllers, and four damaged batteries were due to a late refill of the electrolyte. It has also been noticed that the bad sizing of the generator system and the climatic factors (temperature, humidity) which were not adapted to the good functioning of the batteries cause their breakdown. The most frequently observed breakdown causes of the cables are (**Figure 13(b)**) cable break age and corrosion which lead to short circuits and worn out cables, and great length cables can also cause the voltage drop and energy losses at the end of the system.

7.1.1.2 Causes at the level of the inverter and PV module

Breakdown on inverters is frequently caused by overvoltage (**Figure 14(a)**)). They have as origin the nonfunctioning of charge controllers and the frequent interruption of electricity of the network. For PV modules the most frequent causes

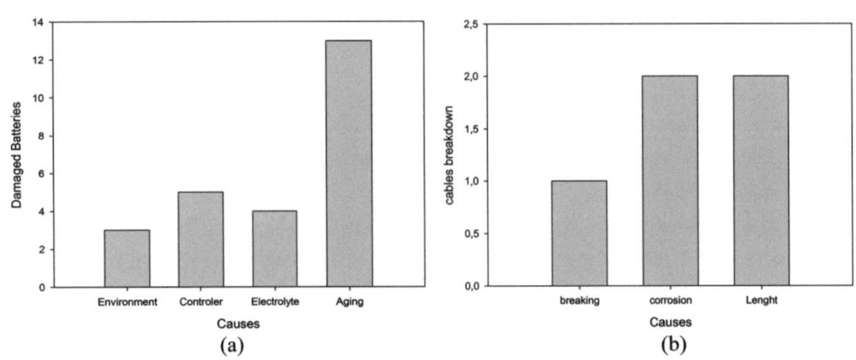

Figure 13.
(a) Damaged batteries and their causes and (b) cables breakdown versus causes.

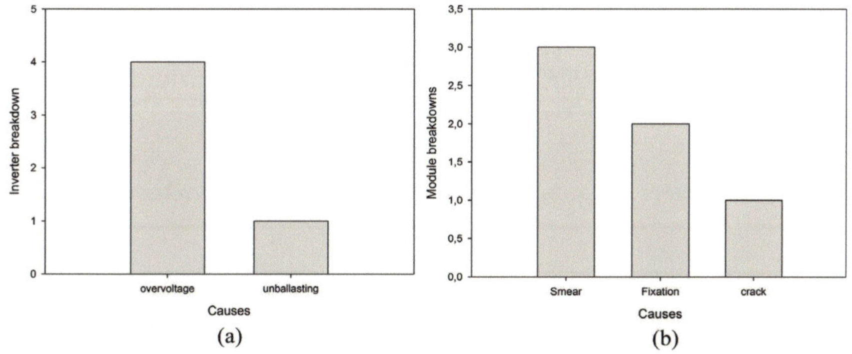

Figure 14.
(a) Inverters faults versus causes of breakdown and (b) PV module breakdowns versus causes.

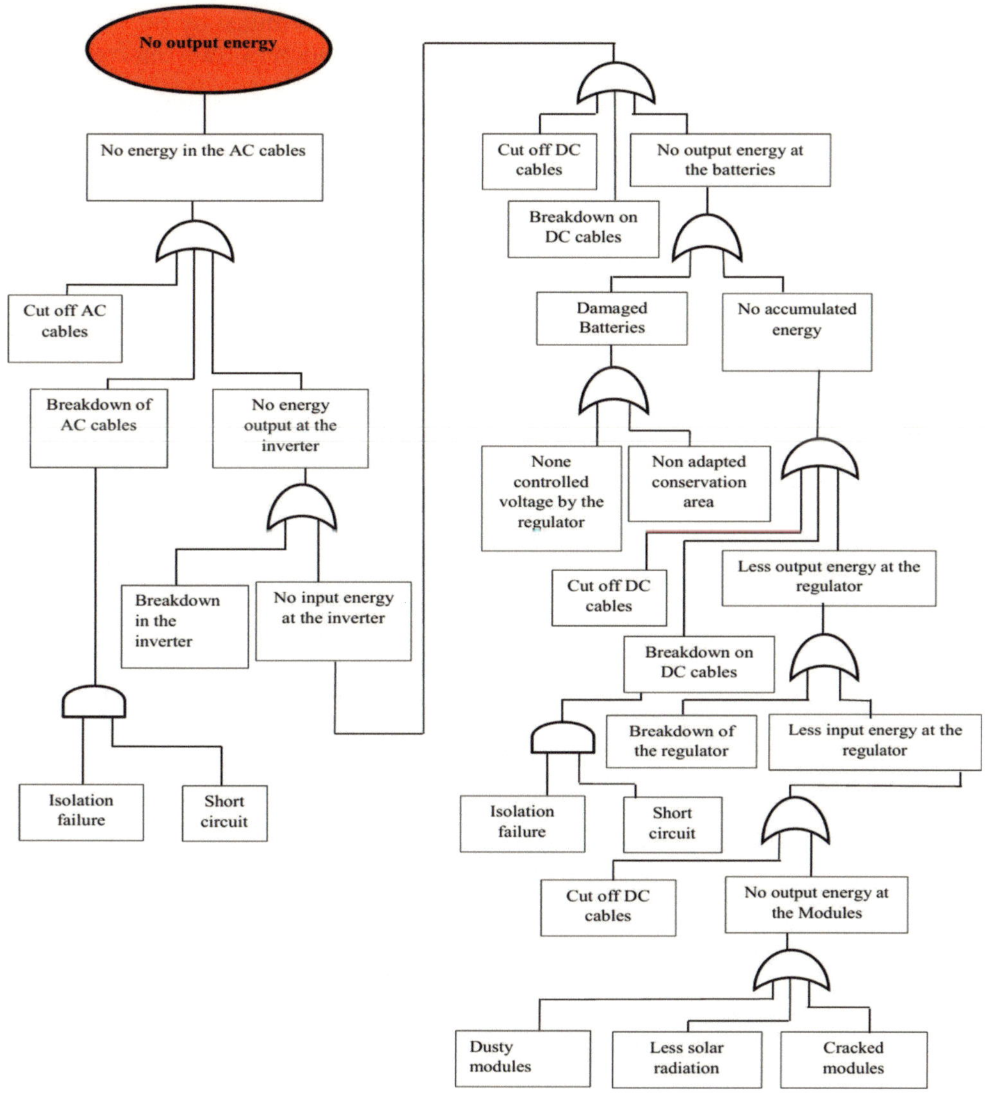

Figure 15.
Breakdown tree diagram for detecting breakdowns in a PV system.

are **Figure 14(b)**: dust deposit, bad fixing on their supports, and accidental cracking. The first two causes which do not lead to the stop of the system decrease the efficiency of the system (output energy), while the third cause leads to the stop of the system which necessarily needs a replacement.

7.1.2 The breakdown tree diagram used for the diagnosis of the breakdown in the studied system

During corrective interventions, most of the workers overcome the breakdown without trying to eliminate their causes or without investigating on the causes of the breakdown. The exploitation of maintenance files shows that many elements were replaced within a short period of time. This situation led to frequent breakdown of the system even though the bad element was replaced. It became very important to put in place a breakdown diagnosis method in order to eliminate breakdown and causes. The breakdown tree technique helps to graphically represent the possible combinations of the events that permit the realization of a non-needed predefined event. The breakdown tree is then made of levels of events linked by gate (initially logic gate). By using this representation and a logical deduction (moving from

Components	Failures	Possible causes	Effects on the system	Observable parameter	Criticality			
					G	O	D	C
Photovoltaic generator (PVG)	$V_{m,PV} < V_{ref,PV}$ T = 25°C E = 800 W/m^2	• Diode bypass short circuit • Defective module in serial	• Decrease of tension	• Voltage	5	2	2	20
	$I_{m,PV} < I_{ref,PV}$ T = 25°C E = 800 W/m^2	• Cutting connection wires • Defective anti-return diodes	• Decrease in current and flow	• Current	3	5	2	30
	$P_{mPV} < P_{refPV}$ T = 25°C E = 800 W/m^2	• Shadow-related design flaw • Dirt of the panels	• Low water flow	• Voltage • Current	5	5	4	100
	$P_{m,PV} = 0$ T = 25°C E = 800 W/m^2	• Corrosion or looseness of the connection terminals • Defective fuse	• No power	• Voltage • Current	9	7	4	252
Inverter	$P_{Inv} = 0$ T = 25°C E = 800 W/m^2	• Inverter failure • Defective power cables or poor tightening at the inverter input	• No water flow	• Input power • Power output	9	7	4	252
Pump	$Q_m = 0$ Not functioning (pumping stop) E = 800 W/m^2	• Clogged strainer • Defective wheels • Lowering of the water level	• No flow of water	• Flow rate	9	7	4	252
Engine	$Q_m = 0$ Not starting E = 800 W/m^2	• Defective phases • Low voltage • Defective mechanical sea	• No flow of water	• Voltage • Current	9	7	4	252

Table 4.
Possible failures on the photovoltaic pumping system according to FMECA.

Components	Failures	Possible causes	Solution
Photovoltaic generator (PVG)	$V_{m,PV} < V_{ref,PV}$ T = 25°C E = 800 W/m^2	• Diode bypass short circuit • Defective module in serial	• Replace the diode • Check and replace the faulty module
	$I_{m,PV} < I_{ref,PV}$ T = 25°C E = 800 W/m^2	• Cutting connection wires • Defective anti-return diodes	• Check and replace the connection wire • Replace the faulty diode
	$P_{mPV} < P_{refPV}$ T = 25°C E = 800 W/m^2	• Shadow-related design flaw • Dirt of the panels	• Clear the objects causing the shadow • Clean the modules
	$P_{m,PV} = 0$ T = 25°C E = 800 W/m^2	• Corrosion or looseness of the connection terminals • Defective fuse	• Tighten or change the connection terminals • Change the fuse
Inverter	$P_{Inv} = 0$ T = 25°C E = 800 W/m^2	• Inverter failure • Defective power cables or poor tightening at the Inverter input	• Replace the inverter • Remove or replace the power cable
Pump	$Q_m = 0$ Not functioning (pumping stop) E = 800 W/m^2	• Clogged strainer • Defective wheels • Lowering of the water level	• Unclog the strainer • Change the wheels • Check the water level is at least 1 m above the suction body of the pump during operation
Engine	$Q_m = 0$ Not starting E = 800 W/m^2	• Defective phases • Low voltage • Defective mechanical sea	• Replace the phases • Change the seals

Table 5.
Their solutions to the possible failures in a photovoltaic pumping system according to FMECA.

effects to causes), it is possible to cast out the causes from the effects, from the non-needed event to base events, independent to one another and probable (**Figure 15**).

7.2 Photovoltaic water pumping systems

In PV water pumping systems, the main objective is to collect data, diagnosing the system and proposing and implementing solutions that will optimize the operation of the system in order to satisfy the water need of the population all over the year. Indeed from the collected and measured data, **Tables 4** and **5** show the

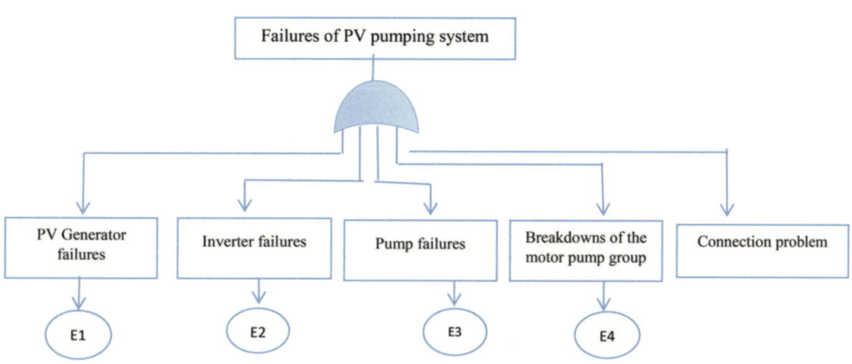

Figure 16.
*Failure of the photovoltaic pumping systems (events E1, E2, E3, and E4 are shown in **Figures 17–20**).*

failures we may encounter in our installation as well as the possible causes and solutions.

To facilitate the procedure of these pumping stations, breakdown tree diagram is constructed in a deductive manner as highlighted in the **Figure 16**. For each event E1, E2, E3 and E4 in the **Figure 16**, breakdown tree diagram for failures encountered are presented respectively in the **Figures 17–20**.

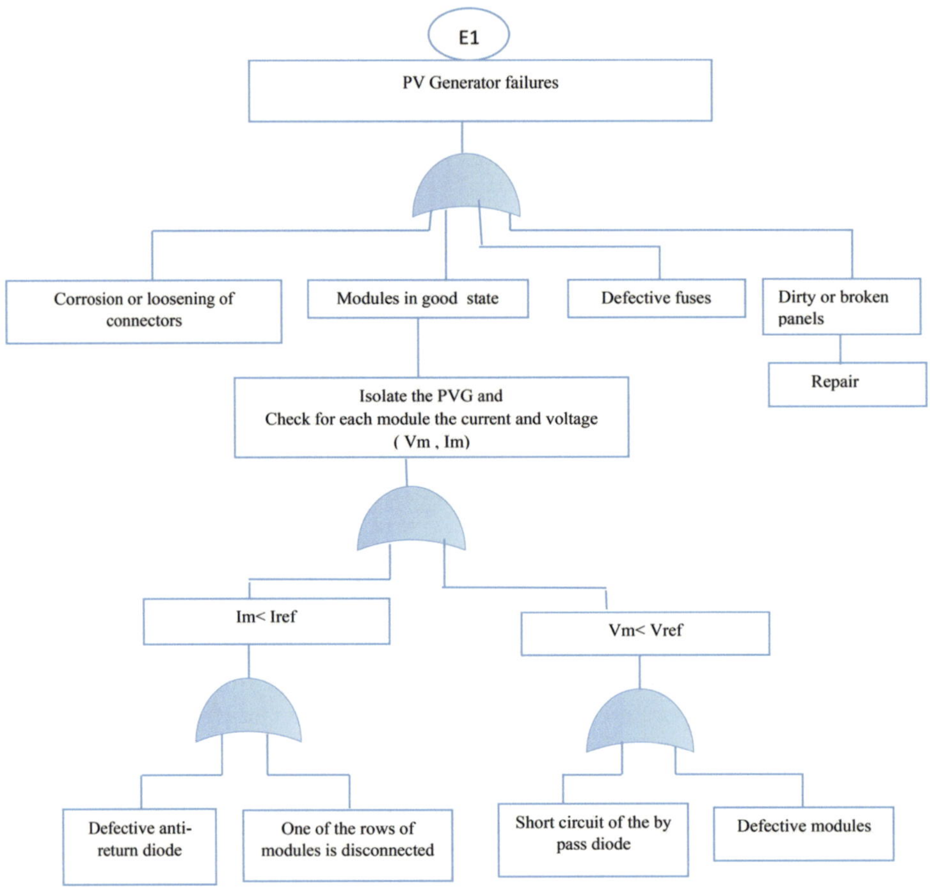

Figure 17.
Photovoltaic generator failure tree.

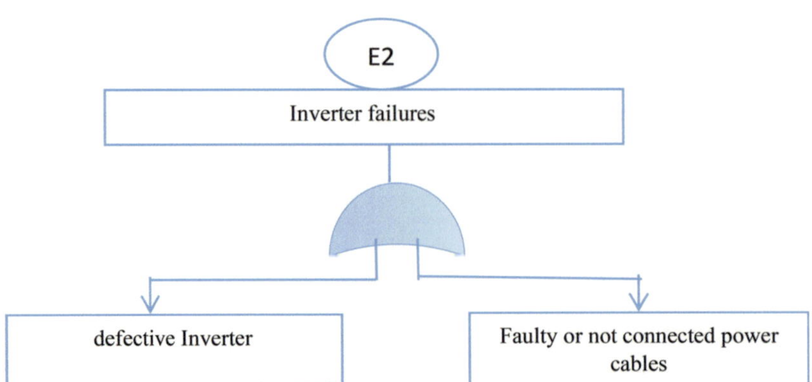

Figure 18.
Inverter failure tree.

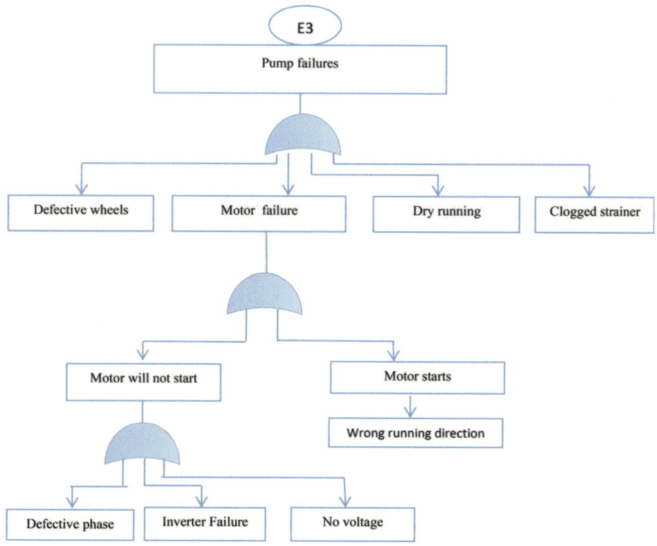

Figure 19.
Pump failure tree.

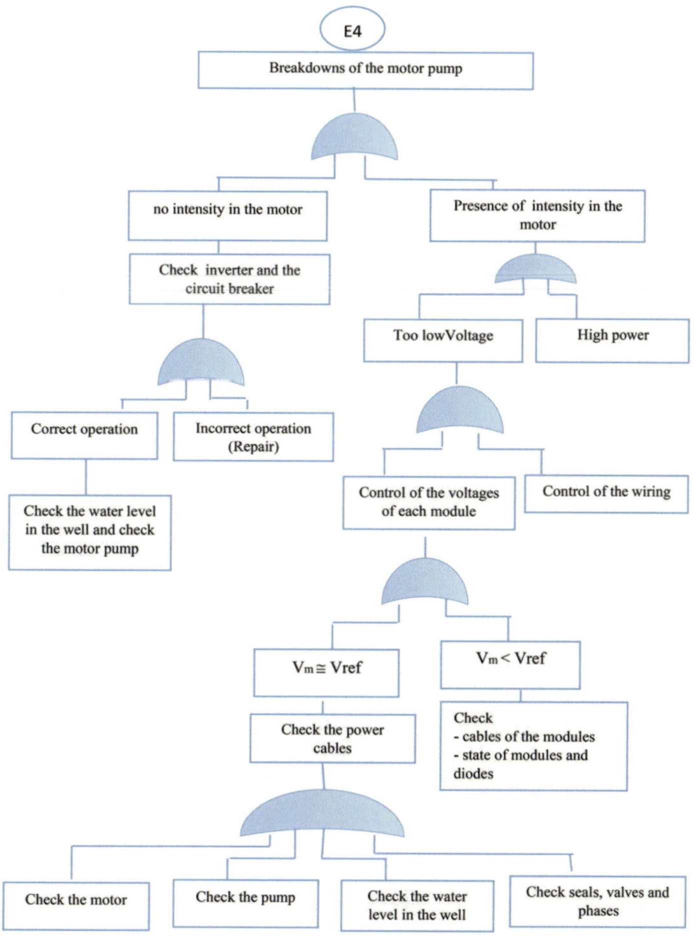

Figure 20.
Pump unit failure diagram.

8. Conclusion

Maintenance operations have a crucial interest in the viability evaluation and the analysis of the life expectancy of a PV system. In this study, we have shown the maintenance techniques which can enable the best diagnosis of breakdown. After the exploitation of maintenance file report data of the twenty backup PV systems and two PV water pumping systems installed in more than 15 towns in Cameroon, it can be concluded that the most vulnerable element of a solar PV system is the battery since this element represents 64.9% of the breakdown recorded. Among the 20 backup PV systems subject of our study, it appears that 50% received their first curative intervention from the 5th year. The FMECA method for PVWPS shows that the criticality of the installation varies from 252 for the inverter, 402 for the PV generator, and 504 for the motor pump. Particular attention must be paid on preventive operations in order to eradicate causes of frequent breakdowns of the components in general and motor pump and batteries in particular. That is why breakdown tree diagram is proposed for the rapid determination of breakdowns on the studied PV systems and the steps or order of detecting breakdowns on a system.

Author details

Kodji Deli* and Djongyang Noel
Department of Renewable Energy, National Advanced School of Engineering, University of Maroua, Maroua, Cameroon

*Address all correspondence to: delikodji27@yahoo.fr

References

[1] Deli K, Tchoffo Houdji E, Djongyang N, Njomo D. Operation and maintenance of back-up photovoltaic systems: An analysis based on a field study in Cameroon. African Journal of Science, Technology, Innovation and Development. 2017;**9**(4):437-448

[2] International Energy Agency (IEA). Statistics, Renewables Information. 2011

[3] International Energy Agency (IEA). Technology Roadmap Solar Photovoltaic Energy. 2014

[4] International Energy Agency. Trends 2013 in Photovoltaic Applications: Survey Report of Selected IEA Countries between 1992 and 2012. Report IEA-PVPS T1-23:2013. 2013

[5] Moreno-Garcia IM et al. Real-time monitoring system for a utility-scale photovoltaic power plant. Sensors. 2016; **16**(6):770

[6] Taylor M, Ralon IA. The power to change: Solar and wind cost reduction potential to 2025. Abu Dhabi, United Arab Emirates: Int Renew Energy Agency (IRENA). 2016. Available from: http://www.irena.org/

[7] Rekioua D, Matagne E. Optimization of Photovoltaic Power Systems. Modelization, Simulation and Control. New York: Springer-Verlag London Limited; 2012. ISBN 978-1-4471-2348-4. DOI: 10.1007/978-1-4471-2403-0

[8] Mboumboue E, Njomo D. Potential contribution of renewables to the improvement of living conditions of poor rural households in developing countries: Cameroon's case study. Renewable and Sustainable Energy Reviews. 2016;**61**:266-279

[9] Ma T, Yang H, Lu L. Solar photovoltaic system modeling and performance prediction. Renewable and Sustainable Energy Reviews. 2014;**36**: 304-315

[10] Marion B, Adelstein J, Boyle K. Performance Parameters for Grid-Connected PV Systems, NREL/CP-520-37358. 2005

[11] Marion B. A method for Modeling the current voltage curve of a PV module for outdoor conditions. Progress in Photovoltaics. 2002;**10**:205-214

[12] Kroposki B, Marion W, King DL, Boyson WE, Kratochvil J. Comparison of module performance characterization methods for energy production. In: Paper No. NREL/TP-520-29245. 2002

[13] Deli K, Djongyang N, Njomo D, Kidmo D. A rainy month performance evaluation of photovoltaic module and performance prediction of the PV water pumping systems under Soudano Sahelian climate conditions. International Journal of Electrical Components and Energy Conversion. 2018;**4**(1):61-71. DOI: 10.11648/j.ijecec.20180401.17

[14] Picault D, Raison B, Bacha S, de la Casa J, Aguilera J. Forecasting photovoltaic array power production subject to mismatch losses. Solar Energy. 2010;**84**:1301-1309

[15] Firth SK, Lomas KJ, Rees SJ. A simple model of PV system performance and its use in fault detection. Solar Energy. 2010;**84**(4):624-635

[16] Mellit A, Tina GM, Kalogirou SA. Fault detection and diagnosis methods for photovoltaic systems: A review. Renewable and Sustainable Energy Reviews. 2018;**91**:1-17

[17] Alam MK, Khan F, Johnson J, Flicker J. A comprehensive review of catastrophic faults in PV arrays: Types, detection, and mitigation techniques. IEEE Journal of Photovoltaics. 2015;**5**(3):982-997

[18] Pearce JM. Photovoltaics—A path to sustainable futures. Futures. 2002;**34**: 663-674

[19] EL-Shimy M. Viability analysis of PV power plants in Egypt. Renewable Energy. 2009;**34**:2187-2196

[20] Sanz-Bobi MA editor. Use, Operation and Maintenance of Renewable Energy Systems. Green Energy and Technology. Cham, Heidelberg, New York, Dordrecht, London: Springer; 2014. DOI: 10.1007/978-3-319-03224-5

[21] Nyman D, Levitt J. Maintenance Planning, Coordination and Scheduling. 2nd ed. New York: Industrial Press; 2010

[22] Hatti M. Operation and Maintenance Methods in Solar Power Plants in Use, Operation and Maintenance of Renewable Energy Systems: Experiences and Future Approaches. Switzerland: Springer International Publishing; 2014. DOI: 10.1007/978-3-319-03224-5

[23] Crespo A, Moreu P, Gómez JF, Parra C, López M. The maintenance management framework: A practical view to maintenance management. Journal of Quality in Maintenance Engineering. 2009;**15**(2):167-178

[24] Tam ASB, Price JWH. A generic maintenance optimization framework. In: 7th Asia Pacific Industrial Engineering and Management Systems Conference, Bangkok, Thailand. 2006

[25] Hassanain MA, Froese TM, Vanier DJ. Framework model for asset maintenance management. Journal of Performance of Constructed Facilities. 2003;**17**:51

[26] Hao Q, Xue Y, Shen W, Jones B, Zhu J. A decision support system for integrating corrective maintenance, preventive maintenance, and condition-based maintenance. Construction Research Congress. 2010. pp. 470-479

[27] Madeti SR, Singh SN. A comprehensive study on different types of faults and detection techniques for solar photovoltaic system. Solar Energy. 2017;**158**:161-185

[28] Triki-Lahiani A, Bennani-Ben AA, Slama-Belkhodja I. Fault detection and monitoring systems for photovoltaic installations: A review. Renewable and Sustainable Energy Reviews. 2018;**82**: 2680-2692. DOI: 10.1016/j.rser.2017.09.101

[29] Davarifar M, Rabhi A, El Hajjaji A. Comprehensive modulation and classification of faults and analysis their effect in DC side of photovoltaic system. Energy and Power Engineering. 2013; **5**(04):230

[30] Colli A. Failure mode and effect analysis for photovoltaic systems. Renewable and Sustainable Energy Reviews. 2015;**50**:804-809

[31] Foley G. Photovoltaic Applications in Rural Areas of the Developing World. Vol. 304. Washington, DC: World Bank Publications; 1995

[32] Labouret A, Villoz M. Solar Photovoltaic Energy, 5th edition, Dunod, English Translation 2010. Dunod, Paris: The Institution of Engineering and Technology; 2009

[33] DOE EERE. Procuring Solar Energy: A Guide for Federal Facility Decision Makers. 2010. Available from: https://www1.eere.energy.gov/solar/pdfs/47854.pdf

[34] Martifer Solar SA. PVTRIN Training course-Troubleshooting Guide. 2011. Available from: https://ec.europa.eu/energy/intelligent/projects/sites/iee-projects/files/projects/documents/pvtrin_troubleshooting_guide_for_installers_en.pdf

[35] IEC 61724, Photovoltaic System Performance Monitoring Guidelines for Measurement, Data Exchange, and Analysis; Switzerland, Geneva. 1998

[36] Limnios N, Ziani R. An algorithm for reducing cut sets in fault-tree analysis. IEEE Transactions on Reliability. 1986;**35**(5):559-562

[37] Limnios N. Fault Trees, Control System Robotics and Manufacturing. Unites States: ISTE; 2007. ISBN 10: 1905209304; ISBN 13: 9781905209309

[38] Bun L. Détection et localisation de défauts dans un système photovoltaïque, Laboratoire Génie électrique G2ELAB, Genoble, Rapport de thèse tel-00647189v1 > 2011. 2011

Chapter 9

Automatic Visual Inspection and Condition-Based Maintenance for Catenary

Yan-guo Wang, Dapeng Xie, Qiang Han, Yi Zhang, Wei Zhou, Xiantang Xue, Wenxuan Zhang and Kai Tao

Abstract

Defects on catenary components are a major part of device faults as a result of a much higher tension on high-speed catenary, such as looseness of bolts, component broken, and component missing. Traditional inspection on catenary components has to be performed only at night with human eyes. Not only the inspection speed is very slow but also the inspection results are not reliable, as a result of the poor lighting environment and long-time working tiredness. In this chapter, we present an automatic visual inspection system for checking the status of components on catenary. A dedicated designed camera system is mounted on an inspection car, which covers almost all the components to be checked and gives great details of each component. Considering the great data storm at each catenary post, high-performance servers with GPU acceleration are used, and technologies of multi-thread and parallel computing are exploited. Furthermore, an intelligent analysis framework is proposed, which uses structural analysis to localize each component in the image and perform automatic detection based on different features such as geometry, texture, and logic rules. The system has been successfully used in China's high-speed railways, which shows great advantages in the catenary inspection application.

Keywords: catenary inspection, machine vision, visual inspection, high-speed railway, security

1. Introduction

Compared with traditional defects on stagger, arc, contact force, etc., defects on catenary components become more and more important on high-speed railways [1]. Defects on catenary components are a major part of device faults as a result of a much higher tension on high-speed catenary, such as looseness of bolts, component broken, and component missing. **Figures 1** and **2** illustrate some typical defects on catenary components.

Traditional inspection for catenary components on high-speed railways has to be performed only at night with human eyes (**Figure 3**). Not only the inspection speed is very slow but also the inspection results are not reliable, as a result of the poor lighting environment and long-time working tiredness.

Figure 1.
Component missing.

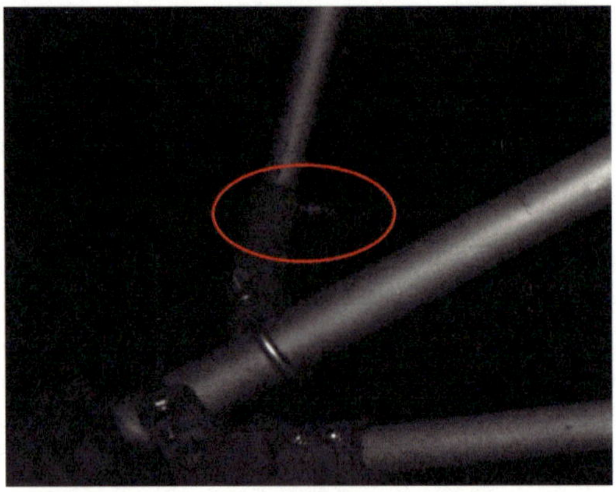

Figure 2.
Looseness of bolts.

Due to the great advances in machine vision technologies [2, 3], some visual inspections or visual measurement systems have been developed in the field of infrastructure inspection on track, catenary, etc. [4–8]. However, the development of a visual inspection system for catenary components is not an easy task. First, catenary is composed of many types of components, and these components are mounted in a large area and vary greatly in size and location, which brings great difficulties in the design of camera system. Second, most components concentrate at catenary posts, which means a great data storm for the image capturing, processing, and saving. Finally, the automatic data analysis is also of great challenge, as a result of the variety and large number of defect types.

In this paper, we present an automatic visual inspection system for checking the status of components on catenary. A dedicated designed camera system is mounted on an inspection car, which covers almost all the components to be checked with an image resolution up to 0.5 mm/pixel. A post detection module is integrated in order to trigger the cameras and lighting device at a proper

Figure 3.
Traditional inspection.

location. Technologies of GPU acceleration, multi-thread and parallel computing, are exploited in the image acquisition, image compression, and data storage. Furthermore, an intelligent analysis framework is proposed to perform the automatic analysis of image data.

The rest of paper is organized as follows. Section 2 gives an overview of the system architecture. Section 3 describes the methodology of each system part in detail. Section 4 reports some application results of the system. Then we draw the conclusions in the last section.

2. System architecture

Catenary is composed of contact wire, cantilever, additional wire, hanging post (in tunnels), etc. (**Figure 4**). As a result, different kinds of catenary components are mounted in a large area and vary greatly in size and location.

Considering that most components concentrate at catenary posts, we use groups of super-high-resolution area-scan cameras to capture image data of each region at a specific distance relative to each catenary post. A post detection module is applied to generate an accurate trigger signal for cameras and lighting device. Finally, image data from different cameras are paralleled processed and saved in a common data sever.

The system architecture is illustrated in **Figure 5**.

Figure 4.
Structure of catenary.

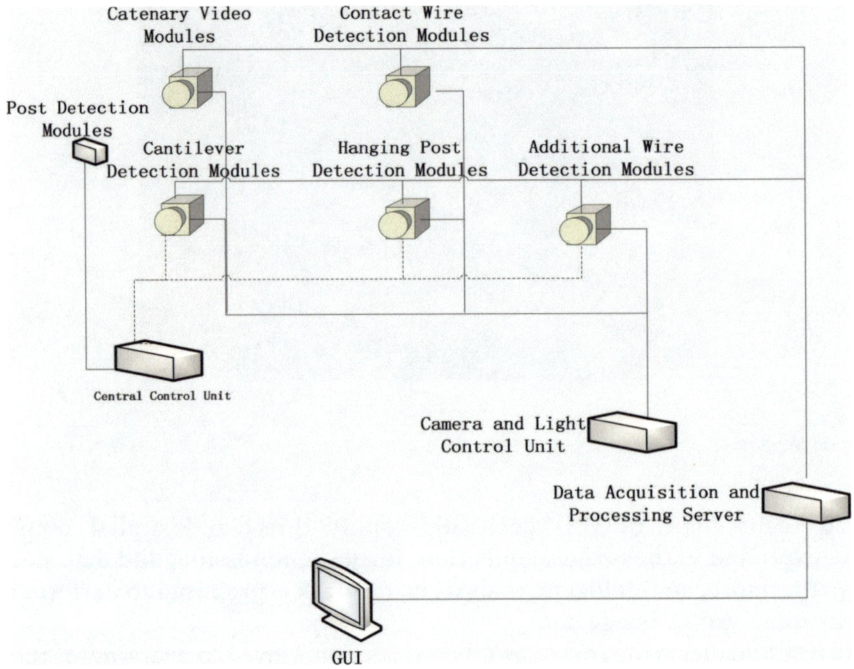

Figure 5.
System architecture.

3. Methodology

3.1 Camera system

Catenary components to be checked are mounted in a large area. It covers [−3500, 3500 mm] in the horizontal direction and [4800, 8100 mm] in the vertical direction relative to the track center. High-resolution image data are needed in order to perform image detection of each component.

In our system, the candidate detection area is divided into several detection regions. A dedicated designed camera system is mounted on an inspection car, which is composed of many area-scan cameras with different directions and fields of view. Super-high-resolution cameras are used which have a resolution up to 29 megapixel.

Table 1 gives details of the camera system. The camera system is carefully designed so that it covers almost all the components to be checked with an image resolution up to 0.5 mm/pixel, which gives great details of each component.

In order to get a brighter light and better image quality at night, groups of strobe light are applied in the system.

Figure 6 illustrates the design for part of the camera system.

3.2 Post detection

The work of the system relies on an accurate generation of trigger signals at a specific distance relative to each catenary post. As a result, a reliable post detection module is very important in the system. The post detection module should be adaptive to different detection environments and types of catenary.

In order to achieve a reliable detection, several laser distance sensors are used which are mounted upward. The range of distance between the sensors and steady

Detection modules	Camera resolution (megapixel)	Number of cameras	View directions
Cantilever	29	14	Forward and backward
Contact wire	25	2	Left and right
Additional wire	16	4	Forward and backward
Hanging post	16	4	Forward and backward
Catenary video	4	1	Forward

Table 1.
Camera system.

Figure 6.
Camera system.

arms is utilized as the criterion in the data analysis of post detection. Measurement data are captured and analyzed in real time. The trigger signals are generated and distributed to each camera and strobe light.

The principle of the post detection module is illustrated in **Figure 7**.

Laser sensors in the 1# group are used when the inspection car runs forward, and those in the 2# group are used when the inspection car runs backward. Finally, an accurate photographic distance at each post can be achieved for each camera as expected.

3.3 Image acquisition, processing, and saving

In the camera system, up to 25 super-high-resolution area-scan cameras are used to capture image data of each catenary component. In the maximum inspection speed of 160 km/h, the overall original image data is up to 10 Gb/s! It is a great challenge for the image acquisition, compression, and data storage.

Considering the great data storm at each post, high-performance servers are used in the system. The architecture of multi-thread processing is shown in **Figure 8**. The tasks of image acquisition, image compression, and image saving are distributed in different threads for each camera and are performed based on the technologies of parallel computing and GPU acceleration.

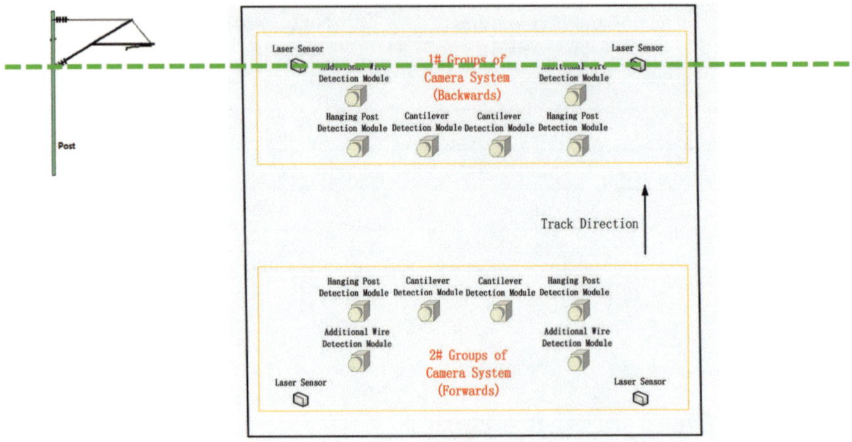

Figure 7.
Post detection.

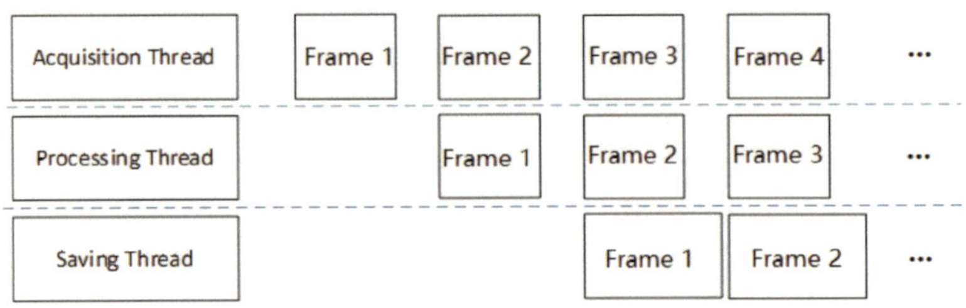

Figure 8.
Multi-thread processing.

Although the huge amount of image data are captured and compressed on several acquisition platforms, the final compressed JPEG images are saved in a common data server, as illustrated in **Figure 9**.

There are labels in the name of each image file, which indicates the camera, post number, detection region, and frame number. The image data are well aligned, and all the image data for the same catenary post are saved in a common file folder, which is helpful for the data management and analysis.

3.4 Image detection

Technologies of visual inspection and image detection have already been used in the detection of track components such as fasteners and track slabs [7, 8]. However, it is quite different for the detection of catenary components. Not only the number of components and defect types to be checked is much larger, but also catenary components vary greatly in size and location, which brings more difficulties in the image detection.

In this paper, we proposed an intelligent analysis framework. Firstly, as shown in **Figures 10** and **11**, structural analysis is applied in order to localize each component in the image. Technologies of matching are exploited here, based on pair wise constraints [9] between the main catenary components such as steady arm, registration arm, cantilever tube, etc.

Secondly, according to the analysis on the characteristics of different components and typical defects, elaborate detection algorithms are separately developed for each component, based on different features such as geometry, texture, and logic rules (**Figures 12–15**).

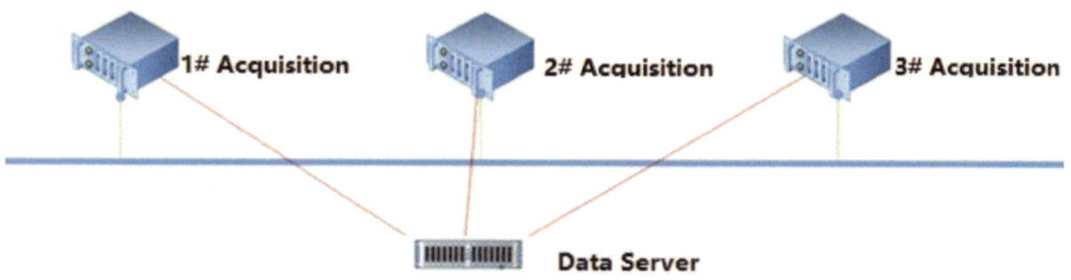

Figure 9.
Data storage.

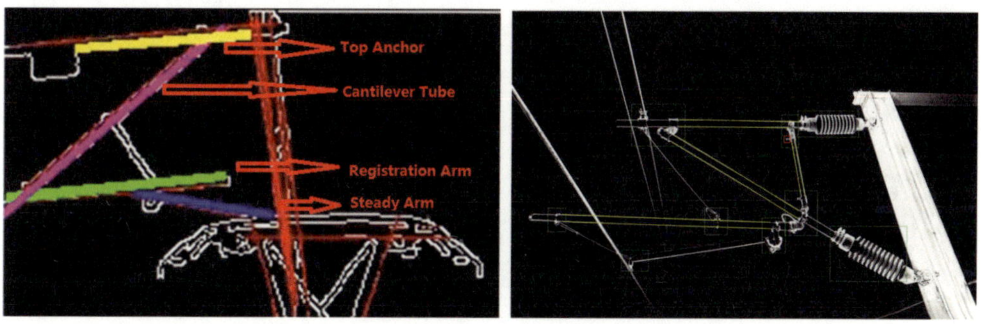

Figure 10.
Structural analysis.

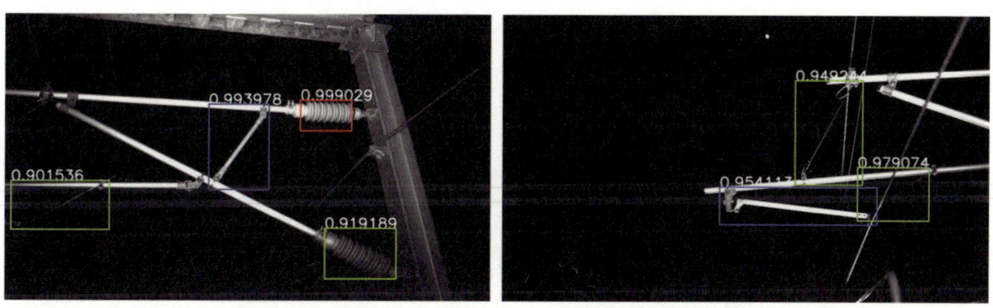

Figure 11.
Structural analysis.

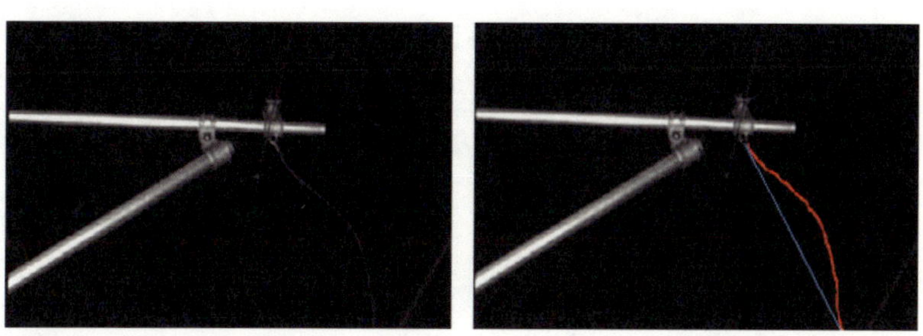

Figure 12.
Detection based on geometry feature.

Finally, only a few of candidate defects are automatically selected and need to be manually rechecked by the system operator, which greatly improves the inspection efficiency (**Figure 16**).

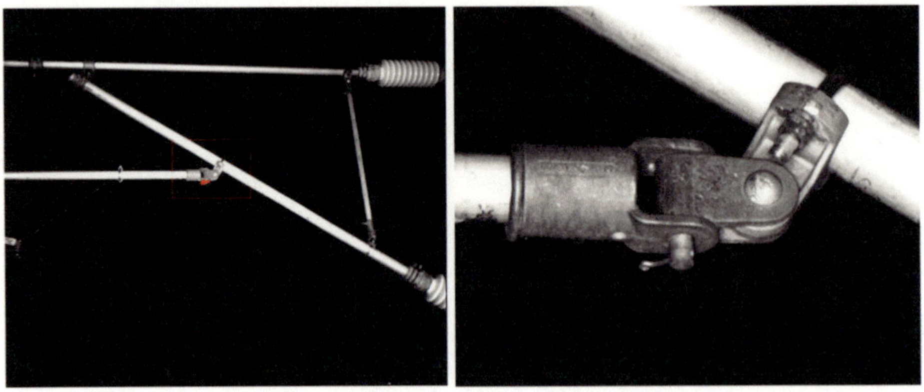

Figure 13.
Detection based on geometry feature.

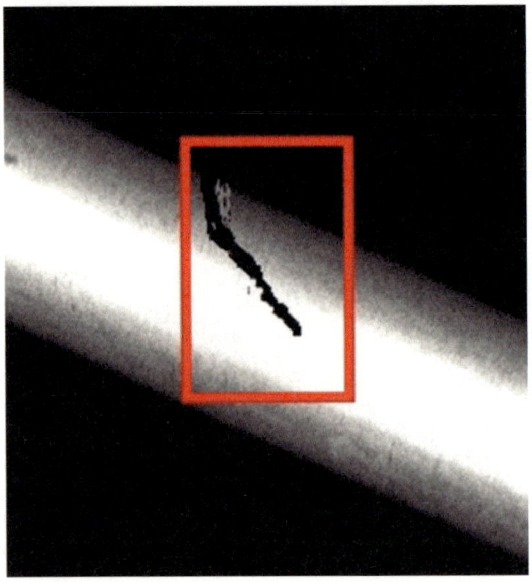

Figure 14.
Detection based on texture feature.

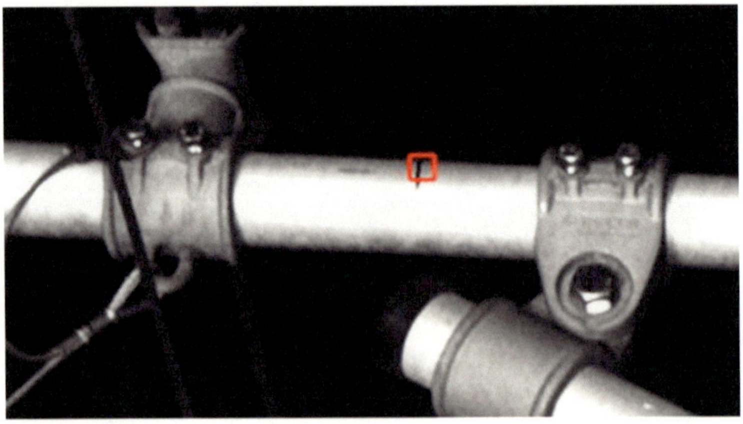

Figure 15.
Detection based on texture feature.

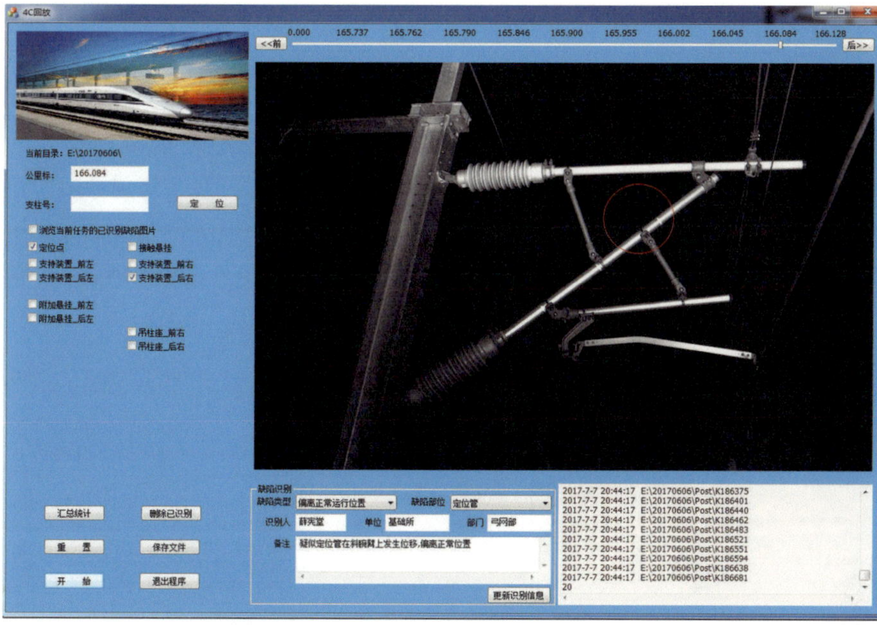

Figure 16.
Manual rechecking for candidate defects.

4. Applications

The proposed system has been installed on several catenary inspection cars which have a maximum inspection speed of 160 km/h. **Figure 17** shows the inspection equipment on one of the inspection car.

The graphic user interface is shown in **Figure 18**. All the camera control and image display of different cameras can be done in the common software, and we can also enlarge a specific part of image to see much detail of any component or region that we are interested in.

Images of different inspection regions are shown in **Figures 19–22**.

The camera system is designed with an image resolution up to 0.5 mm/pixel. **Figures 23–25** show details of some detection regions which are enlarged from the original images. As shown below, the images of catenary components are clear enough to perform the subsequent detection.

Figure 17.
Equipment on the inspection car.

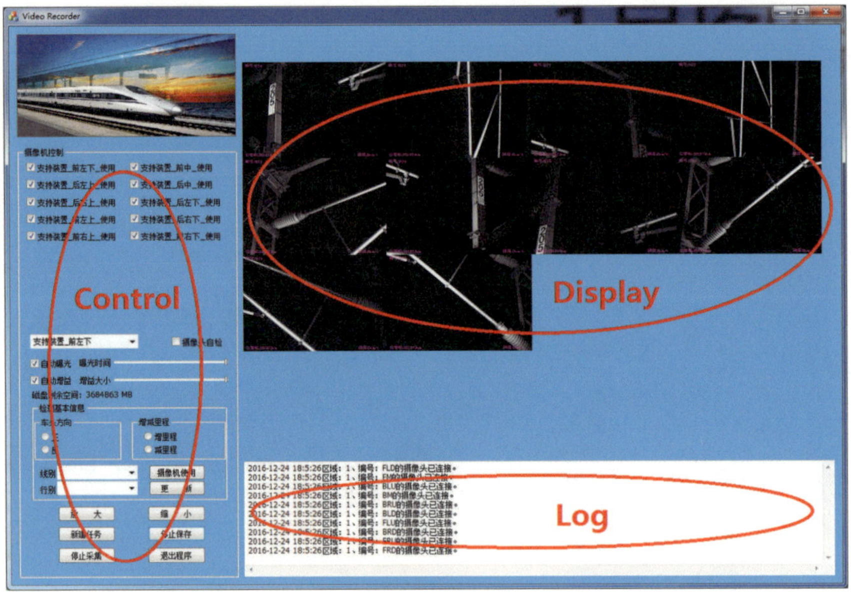

Figure 18.
Graphic user interface.

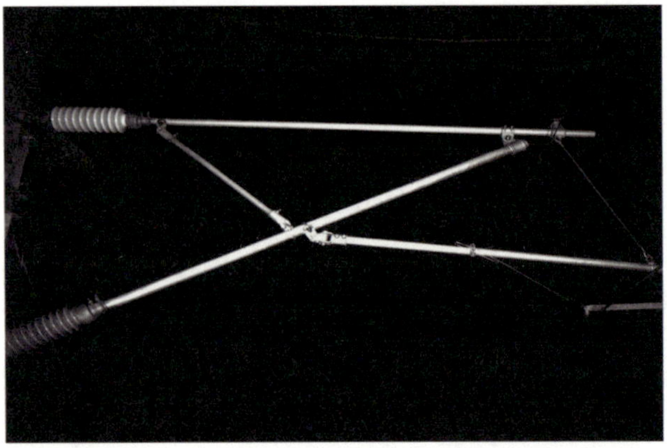

Figure 19.
The whole cantilever region.

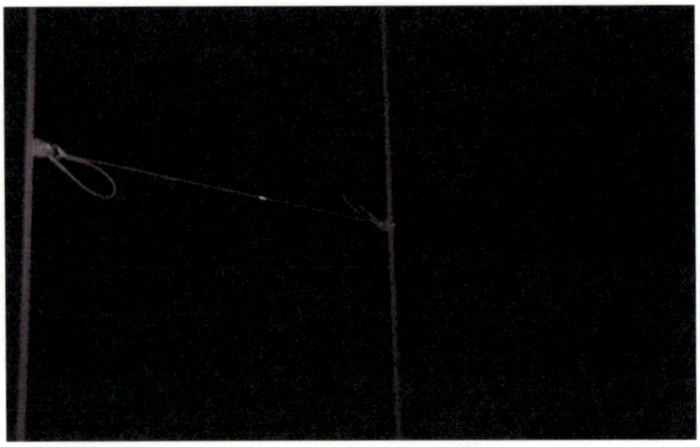

Figure 20.
The contact wire region.

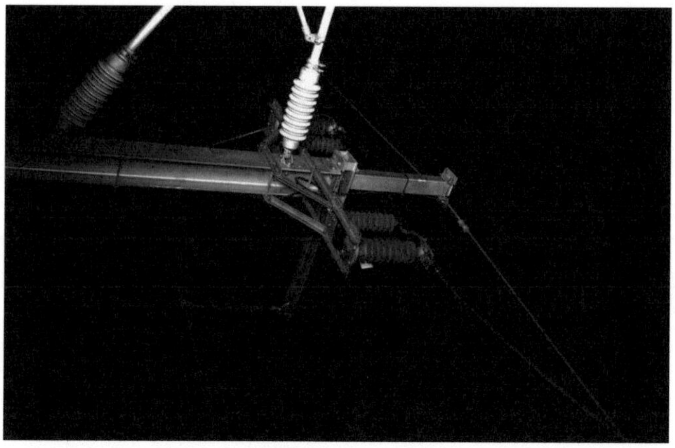

Figure 21.
The additional wire region.

Figure 22.
The hanging post region.

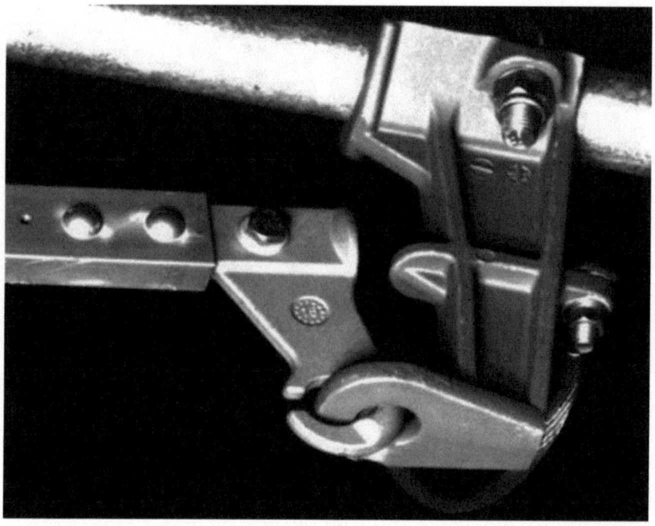

Figure 23.
Enlarged image.

Figure 24.
Enlarged image.

Figure 25.
Enlarged image.

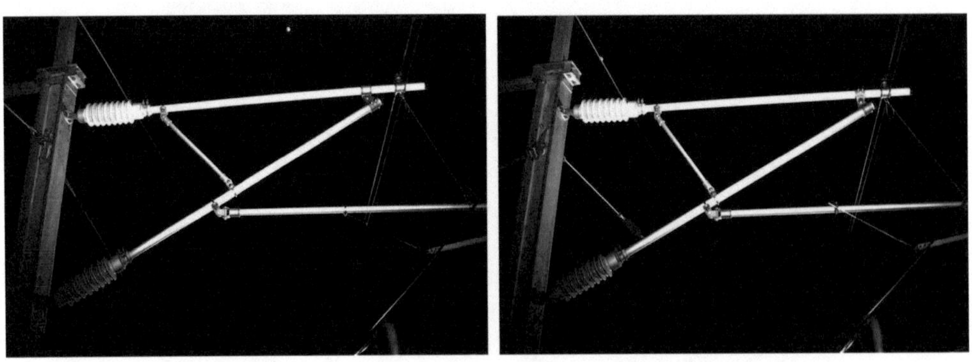

Figure 26.
Images from two inspection runs.

As for the post detection, there are almost no missing detections in the system test. The detection performance is reliable between railway stations. However, a few of false detections exist in some stations, as a result of the complex structure of crossing wires and electric connection wires at stations.

Comparison between images from two inspection runs at the same post is shown in **Figure 26**. We can see the two images are almost the same, which means the accurate post detection and trigger signal generation have been achieved.

Some defects that have been automatically detected by the system are shown in **Figures 27–30**.

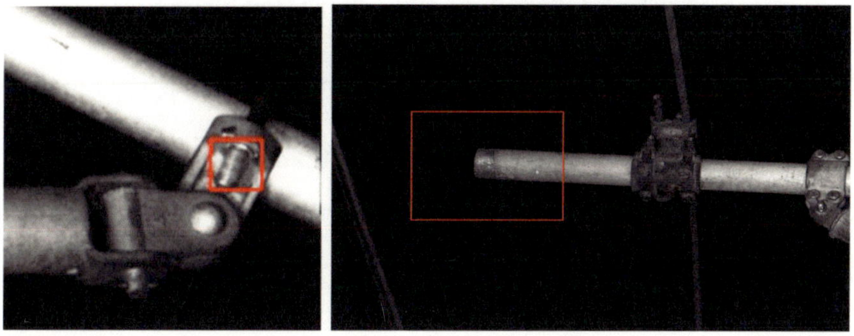

Figure 27.
Component missing.

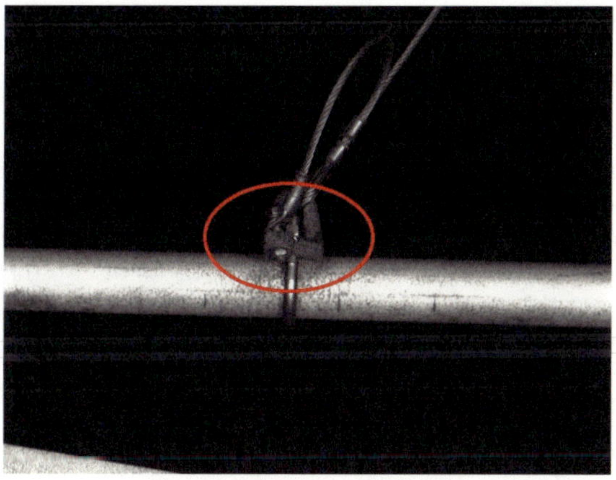

Figure 28.
Looseness of bolts.

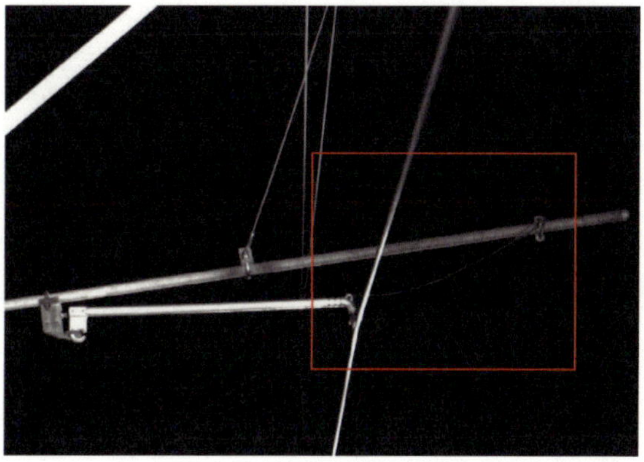

Figure 29.
Relaxation of tightening wires.

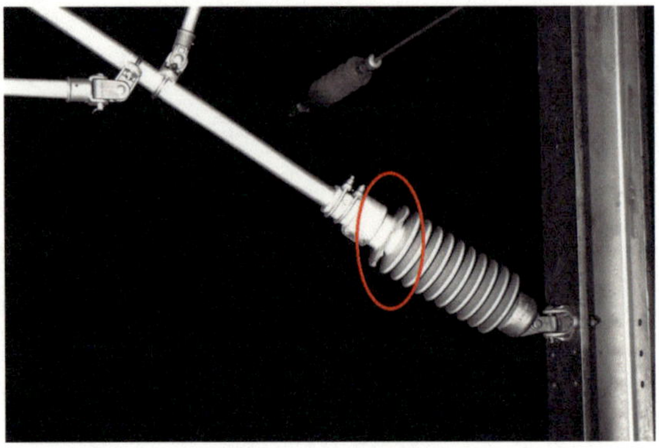

Figure 30.
Broken of insulators.

5. Conclusions

A visual inspection system for catenary components is proposed in this chapter. An image resolution of up to 0.5 mm/pixel is achieved, which covers almost all the components to be checked. An intelligent analysis framework is also proposed, which is used to perform automatic image detection for component defects.

In the past, the component status of catenary is inspected periodically; workers have to go on the railway lines and perform the inspection by human eyes. The infrastructure maintenance will be arranged based on the inspection results. The in-time catenary maintenance is no easy as a result of the poor inspection efficiency. Thanks to the inspection technologies and system proposed in this chapter, the inspection efficiency for component status of catenary is greatly improved, and it is of great use for the catenary maintenance. The status of catenary is well managed based on the automatic visual inspection, and the manner of catenary maintenance is improved from periodically maintenance to condition-based maintenance.

Acknowledgements

This research was partly supported by the National Key Basic Research Program of China (973 Program) (Grant No. 2013CB329406), China Railway Corporation Program on Scientific Research and Development Project (Grant No. P2018G002), and China Academy of Railway Sciences Research Program (Grant No. 2014YJ052).

Author details

Yan-guo Wang*, Dapeng Xie, Qiang Han, Yi Zhang, Wei Zhou, Xiantang Xue, Wenxuan Zhang and Kai Tao
Institute of Infrastructure Inspection, China Academy of Railway Sciences Corporation Limited, Beijing, China

*Address all correspondence to: ygwang@rails.cn

References

[1] Kiessling F, Puschmann R, Schmieder A, et al. Contact lines for electric railways. 3rd ed. Erlangen: Publicis Publishing; 2018

[2] Forsyth DA, Ponce J. Computer Vision: A Modern Approach. 2nd ed. America: Prentice Hall; 2012

[3] Davies ER. Computer and Machine Vision: Theory, Algorithms, Practicalities. 4th ed. Oxford: Elsevier; 2012

[4] Richter U. Contact and non-contact comprehensive catenary inspection system. Railway Technology Review. 2009:8-10

[5] Wang Y-g, Sun Z, Shiping G, et al. Real-time measurement for the angle of steady arm on high-speed railways. In: Proceedings of the 10th World Congress of Railway Research (WCRR'13); November 2013; Sydney

[6] Zhou W, Sun Z, Ren S, et al. Measurement method for geometric parameters of overhead contact line based on multi-view stereovision. China Railway Science. 2015;**36**(5):104-109

[7] Dai P, Wang W, Ren S, et al. Automatic detection technology of broken fastenings using image processing and adaboost learning algorithm. In: Proceedings of the 10th World Congress of Railway Research (WCRR'13); November 2013; Sydney.

[8] Li Y, Trinh H, Haas N, et al. Rail component detection, optimization, and assessment for automatic rail track inspection. IEEE Transactions on Intelligent Transportation Systems. 2014;**15**(2):760-770

[9] Leordeanu M, Hebert M, et al. Beyond local appearance: Category recognition from pairwise interactions of simple features. In: Proceedings of IEEE Conference on Computer Vision and Pattern Recognition (CVPR'07); June 2007

Chapter 10

RFID in Libraries: Automatic Identification and Data Collection Technology for Library Documents

Timoshenko Igor

Abstract

The chapter describes the main features of the use of the radio-frequency identification (RFID) technology in library activities. The technical capabilities of RFID equipment for unique identification of library documents in information systems for various purposes are shown. The questions of development of library technologies due to the full-featured use capabilities of automatic and radio-frequency identification systems in libraries are presented. The possibility of development of library RFID systems in the direction of the concept of Electronic Product Code (EPC) through the modification of the regulatory framework on the basis of existing harmonized standards is considered. It is shown that this approach to the formation of the regulatory framework will create conditions for increasing the availability of RFID technology for libraries. Development of library RFID systems in the direction of the Internet of things (IoT) concept will significantly increase the integration of the traditional library collections, along with electronic documents, in the modern information space. This will increase the availability of printed documents to readers and contribute to development of library technologies towards development of a global integrated system of library and information support for human activities. This will allow libraries themselves to integrate more fully into the global information space, at the next stage of development of information technologies.

Keywords: RFID, radio-frequency identification, ILS, integrated library system, library automation, EPC, electronic product code, internet of things, IoT

1. Introduction

The issues of integration of automated information systems into the surrounding world as well as its effective use for information support have become important. Automatic identification and data capture technologies play a paramount importance in this. These technologies include bar code, biometric, and radio-frequency identification. Bar-code and radio-frequency identification (RFID) is widely used in library information systems of ABIS. Of these, radio-frequency identification has a significant functional advantage over bar code identification. It is radio-frequency identification that has the potential to contribute to development of library technologies. RFID technology is now firmly established in the life of modern libraries.

For library RFID systems, the main accounting object is the document of the library collections, which has a material carrier of information. These documents primarily include printed publications, which now form the basis of the book stock collections in most libraries. The electronic publications increased the availability of information for users and, at the same time, was a serious challenge for the traditional document collections and traditional library technologies. Radio-frequency identification technology allows increased mobility of traditional documents in the electronic information space.

Full-featured use of RFID system capabilities enables development of library technologies through the use of existing experience and the latest achievements in the field of automatic identification and object management.

In these circumstances, the actual task is to understand the experience, fixing it as generally accepted rules and standards, the implementation of which can ensure further development of both RFID technology and library work technology.

2. History of RFID systems in libraries

The first attempts to introduce radio-frequency identification technology in libraries began in the mid of 1990s. A pioneer in this field was "3 M" company (USA), which since the end of 1960s started to produce and install radio-frequency EAS systems in libraries and since the beginning of 1990s started production of radio-frequency identification devices for libraries. 3 M company announced the first project of school library automation, based on RFID, in 1994, and in the 2000s, several thousand libraries in Europe and the United States had already implemented RFID technology. A lot of practical experience has already been accumulated by this time [1].

Most of the projects used RFID equipment of companies such as 3 M, TAGSYS, and FEIG Electronics of 13.56 MHz band. Commercial attractiveness of implemented projects brought to this market a large number of new participants, among which, along with specialized companies, there were many commercial IT companies of a wide profile working with RFID systems in the field of warehouse logistics. This fact can explain the emergence of RFID library projects based on UHF equipment (860–960 MHz), which is an alternative to the equipment of "traditional" library HF band (13.56 MHz).

In Russian libraries, the technology of radio-frequency identification appeared somewhat later [2]. First, the RFID library project was implemented by "ANTIVOR" company on the basis of Russian ILS "IRBIS64" in the library of the graduate management school of St. Petersburg state University, in 2007 [3].

The first full-featured project implemented in a large Russian library on the basis of Russian RFID-equipment was a research project implemented in 2008–2009 by non-commercial partnership "International Center of Technology Transfer " (NP "ICTT"), in cooperation with the "GPNTB of Russia" library [4]. The project used the equipment of HF band, appropriate for library conditions and is claimed in foreign libraries.

3. Regulatory framework of RFID library systems

Widespread use of RFID systems in libraries required systematization of the acquired experience. The first library standard for application of RFID technology appeared in Denmark in 2005 [5]. This standard has been directly supported by many countries of the world, and in 2011 Technical Committee ISO/TC 46/SC 4 has established a system of international standards ISO 28560, representing a group of three standards, under the general title "Information and Documentation. RFID in Libraries." Adopted standards defined the main technical parameters of library RFID systems, as well as the structure and protocols of data exchange with library automation systems. Later, in 2014, the system of standards was revised and a fourth part was added to it. The new standard part was defining the use of UHF band RFID equipment (850–960 MHz) in libraries.

Currently, the international standard ISO 28560 set consists of four parts. To date, all four parts of the standards system have been introduced into the Russian standardization system as identical to international standards.

ISO 29560-1 [6] standard defined data elements used in the cataloging of documents in library collection, which can be placed in the memory of RFID tags and is used for automation of technology operations in libraries. The entire set of 26 items is given, of which only two items are required—the "Primary Item Identifier" and the "Owner Institution (ISIL)". The ISIL code is defined in the ISO 15511 [7] standard. Both identifiers represent in the aggregate the "International Library Item Identifier" (ILIL), defined in ISO 20247 [8]. Libraries are invited to decide of specified elements what should be used, based on the needs and capabilities of their automation system. Data elements are presented without specifying the location in the label memory. In general cases, labels may have different organization for different types. There are also no defined encoding conditions in which data can be represented. These conditions are defined in the following parts of the standard.

ISO 29560-2 [9] standard defines the way of placing of data elements, defined in the first part, in the tag memory, based on standard rules of coding of the object identifier structure defined in ISO/IEC 15962. Standard data element placement rules allow for flexible encoding of variable-length data and different formats. Their application allows the use of RFID tag resources with the greatest efficiency. The data encoding rules in this standard can be applied for any type of labels.

ISO 29560-3 [10] standard is based on the principles set out in the Danish national standard and on the experience of its use in other countries. Data structures presented in the standard are focused primarily for HF (13.56 MHz) band type labels conforming to ISO/IEC 18000-3 [11] Mode 1 standard. Such type of tags includes NXP company labels of ICode SliX specifications. User memory of these labels has a capacity of 112 bits and is divided into 28 blocks of 32 bits available for reading and writing by special commands of RFID reader. Other types of radio-frequency labels are considered in terms of their compatibility with the base type.

Data element allocation principles, defined in ISO 29560-3 standard, are based on a fixed data structure consisting of several blocks. In total, four types of data blocks are defined, of which only the "Basic block" is obligatory for programming, which is a rigid structure, consisting of fixed length fields.

The "Basic block" contains data elements, defined in the first part of the standard as mandatory:

- Primary item identifier
- Owner institution (ISIL)

as well as data elements defined in the first part of the standard as optional:

- Type of usage
- Set information

which have acquired the status of mandatory under this standard.

Additional "Structured Extension Blocks" are used to store data elements from full set that are not included in the "Basic block". The standard defines 5 types of structured blocks of which the formats are determined by their different purposes in the technological system of the library.

Data allocation principles, defined in the third part of the standard, are not compatible with the rules set out in the second part and they are more stringent. Data compression algorithms are not used in coding; different data elements can be represented in different codes. In general, it can be said that data coding based on the rules of the third part of the standard is less rational than the rules presented in the second part. Adoption of this standard is due to the fact that coding based on the rules of the Danish model became a de facto international standard for libraries long before, and such an international standard was adopted by ISO TC46/SC4 Technical Committee. A large number of libraries in many countries around the world use RFID equipment of HF range, and a huge number of documents were marked with labels encoded according to the rules of the Danish data model. Change to other label types and encoding methods is currently a challenging practice task. This situation is supported by main manufacturers of specialized library equipment. Using UHF RFID equipment in libraries is not popular now, despite the significant advantage of UHF technology in "non-library" areas related to logistics.

The fourth part of the standard (ISO 29560-4 [12]) appeared later than previous three parts and was adopted in 2014 only. The standard defines the rules for placement of data elements presented in the first part of the standard, consistent with coding rules defined in the second part. This part of the standard has been added to allow selection of different frequency bands of RFID equipment, between HF (13.56 MHz) defined in part three of the standard and UHF (850–960 MHz), conforming to ISO/IEC 18000-63 [13], for libraries. Data structures presented in the standard are focused on RFID tags having a block memory organization defined in the EPC global Inc. standard as "Class 1 Generation 2" (EPC C1g2). [14].

The logical memory structure of the radio-frequency labels defined in the fourth part of the standard consists of four blocks, of which only two are available for reading and writing library data elements: "01" (EPC memory) and "11" (User Memory).

For EPC memory block the standard defines the possibility of recording a Unique Item Identifier (UII), composed of the "Primary Item Identifier", the "Application Family Identifier" (AFI), and, all or selectively, two data elements: the "Owner Institution (ISIL)" and the "Set Information".

These data elements in various combinations occupy the entire memory block, and the format of their record does not correspond to the format of the standard EPC code. The presence in memory of a "Unique Item Identifier", in non-EPC format, is determined by the value of a fixed bit in the memory block (bit 17hex = 1), located directly in front of the AFI byte area.

For user memory block, the standard defines the ability to write a set of optional data elements, which is a subset of the set defined in the first part of the standard. The choice of data elements to be written to memory can be arbitrary from a given set and is determined by technological needs of the library.

In general, we can say that the fourth part of the standard defines coding rules applicable to labels with a memory structure corresponding to the EPC "C1g2" specification. In this case, the structure of the "Unique Item Identifier" placed in the EPC memory area is not compatible with EPC code format. This makes library RFID systems based on the fourth part of the standard alternative to EPC systems.

Emergence of ISO 28560 systems of international standards was an important step towards the development of RFID library systems. At the same time, while analyzing the content of the standard, it should be noted that its existing parts are not fully consistent with each other, which is a consequence of the historical situation of the use of RFID technology in libraries. Coding principles defined in the second part are not compatible with coding principles presented in the third part. The third and fourth parts of the standard describe incompatible systems. This inconsistency creates difficulties in the further development of RFID technology in libraries.

4. Principles of identification for library items in the library RFID systems

RFID systems use a unique numeric code stored in the memory of radio-frequency tags as an identifier. The degree of uniqueness of the code is determined by the functional needs of the automated systems in which it is used. The main requirement for the identification of code formation is its uniqueness within the boundaries of a specific system.

In the first projects of library RFID systems, a UID code of the radio-frequency label was often used as a unique identifier of accounting objects. The use of this type of identifiers in the library automation system can only ensure their uniqueness. The UID value is set during chip manufacturing; it is constant, and its structure is determined by needs of radio-frequency label manufacturers. In addition, RFID systems based on UID have significant limitations associated with existing library technology: the impossibility of identification of group accounting items in the case of application of non-inventory registration technology for documents and in the case of accounting of document sets of the book stock collections.

The use of radio-frequency labels in the RFID library system, compatible with ISO 28560 standards, involves the use of a rewritable memory area of the label to accommodate structured data, which includes data elements defined in the ISO 28560-1 standard. One of the mandatory data elements is the "Primary Item Identifier" unique for each document instance in the collection of one library. An arbitrary value, that meets the requirements of the ILS, can be assigned for this element. In this case, it is possible to identify the RFID system of group accounting items, such as the publication, as well as sets of documents. The mandatory data block is supplemented by the "Set Information" element for support of accounting of document set. The block is present by the structure of "total set/part number" elements. In addition, if document identifier is located in the rewritable memory, it becomes possible to structure it in order to support functionality of the general system by means of the RFID system.

The ISO 28560 standard defines the length of the primary item identifier as 16 bytes. If you use one byte to display a single character, you can number 10000000000 (10 quadrillion) instances of documents with direct decimal numbering. If you use alphabetic characters to form an identifier, this number of unique combinations will be much greater. Libraries with such collections of printed publications currently do not exist, and in the foreseeable future their appearance is not expected. This code space redundancy can be used to place additional information in the ID code. It can be used to extend functionality of the RFID library system. The primary item identifier must be a structure, each element of which provides a unique identification of the section of the library collection on its hierarchical level. All elements together make up the code of the primary item identifier, which must be unique within the library collection.

Entering into the primary item identifier of additional data elements is suitable for RFID systems that support automation of technological processes related to inventory, with varying degrees of autonomy from library OPAC. These data encoding method give an additional value to the identifier, and it can complement the standard method for writing data elements to the radio-frequency label memory, as defined in ISO 28560-3, 4. Such encoding methods can get faster response of the RFID system by reducing and simplifying read operations for user memory. Also it can be useful, for example, in the case of use of RFID system equipment or ready-made third-party software modules that do not fully implement functions of data elements encoding, according to ISO 28560.

It is advisable to choose data items, which are used in automation of technological operations by means of the RFID system, for encoding the primary item identifier. In addition, selected data items must also remain unchanged, because the primary item identifier must remain the same throughout the life of the document in the library collection. The "Book number" and the "Set information" data items can be used for automated verification of the book stock collections.

The book number data item indicates document location in the library store. Inserting these data item into the code structure of the primary item identifier may be efficient, if it is unchanging for document and based on unchanging classification characteristics. For example, in the case of semantic arrangement of the book stock collections, it can be compiled on the basis of library classification tables (indices UDCC, DCD, etc.) or on the basis of library collection identifier classification (ISCI defined in ISO 27730 standard [15]). In the case of formal arrangement, such features may be the book format (size and accession arrangement), document type, author number, year of publication, etc.

The book number is used in library processes, related to automated inventory, but it can also be used in other processes, for example, in document pre-ordering systems and to automatically determine possible delivery time of ordered documents from storage location to issuing location.

Entering book index into the structure of the primary item identifier is available for the collection of a separate library, since different libraries may apply different systems of collection arrangement, and formats of book index can vary.

The primary item identifier provides unique codes only within the local integrated library system. To ensure the uniqueness of several libraries, the standard defines an additional data item—"Owner Institution (ISIL)". The ISIL code is the International Standard Identifier for Libraries and related organizations. Its format is defined in ISO 15511 standard as a data structure that consists of ISO 3166-1 country code (alpha-2 type) [16] and organization identifier as an alphanumeric element that identifies library in the national identification system.

The procedure for using the ISIL code to identify library documents is defined in the international standard ISO 20247. This standard defines the International Library Item Identifier (ILII) as a structure, consisting of two elements:

- ISIL or ISCI identifier
- Local item identifier

The ISCI specified as a possible element of the ILII structure is a standard collection identifier and it is defined in ISO 27730 standard. Structurally, the ISCI represents ISIL code with extension in the form of a supplementary collection identifier.

Collection is defined in the standard as a logical group of one or more resources. Collections can also be logically or physically grouped or separated, i.e. a collection can be part of one or more other collections and/or can consist of one or more sub-collections. Collection can be an archive reading room, a digital collection of electronic resources, or OPAC of the library. Collection may consist of documents, combined on a semantic basis and located in different physical sections of the book stock store (in accordance with the type of arrangement adopted in the library) or in different sections of the virtual repository for electronic document collections. The need to use the ISCI collection identifier in the RFID system is entirely determined by the configuration of the technological system of a particular library. If you want to store it in the label data structure, the part of code that extends the ISIL code can be written to the internal code field of the Alternative Owner Institution defined as a data element in the ISO 28560-1.

In general, data structure, presented in the international standard ISO 20247, defines the method of forming identifier of the library item, which provides its unique identification on the scale of several libraries and several countries.

5. Unique identification of library documents in radio-frequency identification systems for various purposes

Automated identification of library document participating in a particular technological operation involves reading data from the tag memory located in the RFID reader working area. At the same time, radio-frequency labels of the same type as the library ones, but not those, can fall into the reading zone. Besides, if library documents are borrowed (or documents are transferred to another library through the interlibrary lending system), they could fall into the reading area of non-library RFID systems for various purposes using the same type of radio frequency tags outside the library. Unauthorized radio-frequency tags in the working area of the RFID system may reduce performance or interfere with the normal operation of the system, for example, to cause malfunction of the accounting system of material objects or to cause false triggering of the system that performs anti-theft functions.

To implement the mechanism of radio-frequency tag selection of the same type in the working area of RFID reader to minimize time of data transmission and exclusion from data exchange of tags that are not included in the system, the Application Family Identifier (AFI) is used.

The AFI is specified by a one-byte code, which is often found in the system memory of the RFID tag. The values of the AFI for various RFID applications are defined by ISO/IEC 15961-2 standard [17]. The hexadecimal "C2h" value is defined for use in libraries. The specified value must be assigned to radio-frequency label of library document located in the area of RFID reading systems for various applications outside library. In this case, they will be ignored or, if necessary, identified as library documents. The "C2h" value can be assigned to the label as a permanent one, at the stage of marking library document, or assigned at the registration of issuing document to reader or in the Interlibrary loan (ILL) system. In this case, when registering return of document, the AFI can be assigned as the "07h" value ("in storage" as defined in ISO/IEC 15961-3 [18]), and it can be used in the RFID library system to implement electronic article surveillance (EAS) functions.

To implement the selection mechanism of the same type of library radio-frequency labels in the RFID reader working area with different data encoding, the Data Storage Format Identifier (DSFID) is used. The DSFID value must be assigned a label at the stage of marking the library document and remain unchanged for the entire period of use of the data recorded in the memory of the label. DSFID values for use in library radio-frequency identification systems are defined in the ISO/IEC 15961-2 standard as follows:

- “06h” for tags encoded according to ISO 28560-2.
- “3Eh” for tags encoded according to ISO 28560-2.
- “1Eh” and “5Eh” values can be used for migration from radio-frequency tags that do not meet the requirements of ISO 28560 standards.

Document of library collection, labeled by RFID tags, may be subject of accounting in the technological system of not the library assignment, for example, in the accounting system in the warehouse of printing house or warehouses, and as a part of transport units in the logistics system for delivery of documents to warehouses of trading organizations or libraries. In addition, documents circulating in the ILL system can be identified in automated mail service systems. For enabling the use of labeled library documents in automated RFID systems, non-library application, data structures written to the label memory must be correctly interpreted by all systems. This possibility is achieved through harmonization of standards, governing the data exchange in systems of different applications.

In library RFID systems that are compatible with the set of standard ISO 28560, for compatibility of library systems with systems of global supply the “Identifier of a trade item GS1” data element can be used, which is optional and is placed in the additional block of the tag data structure, encoded according to ISO 28560-3 rules. Specified data element may contain the Global Trade Item Number (GTIN) [19], assigned by GS1 organization to identify products in the supply chain, which is part of the EPC code system. Unfortunately, encodings provided in ISO 28560 standards are not included in the EPC set of standards currently. Thus, radio-frequency labels of library documents cannot be identified in automated EPC systems operating within the existing standards.

6. Unique identification of library documents in the EPC identification system

Now the global technology of contactless identification on the basis of shaped coding and radio-frequency identification in the world applied for marking of goods and transport units exists and develops. The contactless identification is a basis of automation of account at promotion of production from producer to consumer in the systems of warehouse, transport logistics, and trade. Technology of contactless identification represents a set of compatible technologies developed under the general name of electronic product code (EPC).

The concept of EPC was proposed and developed in the early 2000s by the specialized scientific center for automatic identification “Auto-ID Center” created on the basis of the Massachusetts Institute of technology. Later it was published by the international organization “EPC global, Inc.”. The very name “EPC” is a trademark of this organization. Currently, the concept of the EPC is developed by international organization GS1, which has its offices in a number of countries.

The EPC is a numeric identifier unique to each material object to be accounted. Currently, the most commonly used standard code types are of 64-bit and 96-bit lengths. There is also a 198-bit code standard, and a 256-bit code standard is being developed. The total length of code determines the possible length of data fields and, as a consequence, the width of the code space and freedom of choice of data presentation formats.

All information about objects identified by EPC code is available to organizations within a single Global Data Synchronization Network (GDSN), which allows obtaining data about identified objects by their EPC codes. Currently, there is a migration of data from the GDSN network to the Trusted Source of Data (TSM) network, which is a project of the GS1 organization and is designed for two-way exchange of information for users. The GDSN network acts as a source of data for TSD network aggregators.

The EPC concept is supported by manufacturers of RFID equipment. Currently, market offers a wide range of equipment—radio-frequency tags and readers operating in the UHF range (850–960 MHz). Similar HF band equipment (13.56 MHz) is poorly represented, although it is supported by normative documents at the level of international standards [20]. The technical ability of using such equipment in RFID library systems is defined in ISO 28560-4. Use of the same type of equipment in the systems of warehouse, transport logistics, and library RFID systems, with the possibility of modification of identification codes within the existing standards, gives the principal opportunity to integrate RFID library systems with existing, as well as emerging in various fields, EPC identification systems, combined in the TSD network. This integration will allow multiple uses of a single radio-frequency label at all stages of the movement of printed documents, from their production in the printing house to the end user, not only through trade organizations, but also through libraries. The ability to exchange information about library documents in the TSD network with other participants in this process can significantly expand functionality of library automation systems based on radio-frequency identification. For example, such integration is allowed to improve and automate processes of increasing library stock collections, as well as the processes of document exchange in the ILL system, etc. To implement this possibility, it is necessary to harmonize the standards governing the formats of data presentation in library RFID systems with the set of EPC standards.

The format of the structure of data recorded when marking a library document in the memory of an EPC type label is defined in ISO 28560-4 standard. This format is based on the data of elements, defined in ISO 28560-1, and on the coding principles, defined in ISO 28560-2, with significant differences from the encodings, included in the EPC set of standards, and is not supported by EPC systems.

For RFID systems, compatible with EPC standards of the Serial Global Trade Item Number (SGTIN) system is applied [21]. The SGTIN code is a data structure that corresponds to the general structure of the EPC and consists of a standard heading and the following three elements:

- “Company prefix”—the company identifier of the vendor/owner in the GTIN format is assigned by the GS1 organization. Format is incompatible with the similar in purpose ISIL identifier used in the libraries.

- “Product code” - a generic product identifier in the GTIN format.

- “Serial number” - identifies a specific instance of the product.

First two code elements uniquely conform to standard GTIN code, used in bar coding, and replace elements of the codes EAN and UPC previously used in Europe and the United States. Code GTIN may have a standard length of 8, 12, or 13 characters. The international standard numbers for books (ISBN) and periodicals (ISSN) could be submitted in code GTIN-13 since 2007. For them, the dedicated codes in the table of regional GS1 codes are applied:

- 977—periodicals (ISSN)
- 978 and 979–books (ISBN)

According to ISO 28560-4 standard, when publication is marked with an EPC type radio-frequency tag in the early stages of the supply chain dedicated, the entire "01" tag memory block (EPC memory) must be overwritten by the library value unique identifier of the accounting item with AFI byte. Information about EPC code is lost, and thus, the label ceases to be recognized in EPC systems. The possibility of its return to such an automated system, for example, when delivered to recipient through a transport company or a postal service or when it enters the sales network, is associated with the need to restore the EPC code in its memory, which will lead to the loss of "library" information.

For implementation of this possibility for printed publications, it is possible to mark them, at the initial stage of the supply chain, when printing, by radio-frequency EPC tags with SGTIN code. In this case, the "company prefix" and "product code" values can contain a standard ISBN or ISSN code and are assigned to the label by publisher. The serial number value is also assigned by the publisher and can be a data structure defined locally in the printing house. Assigned values can be used for automation of technological processes, as well as for transportation and storage of publications. Serial number field can be reassigned when such documents arrive to the library. In this case, the serial number can also be structured in accordance with the technological requirements of library cataloging.

To illustrate the possibility of placing the library structure of data elements defined in ISO 28560-1 and ISO 28560-2 in the EPC SGTIN format, the following calculations can be made. The total length of the reassigned "serial number" field for different types of EPC labels can be from 36 to 180 bits. This field can contain data elements that make up unique identifier of accounting object (UII) according to ISO 28560-4, which includes the following elements:

- Primary item identifier—16 bytes
- Set information–2 bytes
- Owner institution (ISIL)—11 bytes

Of these elements, the ISIL code can be placed in a user memory block, so the total length of UII without AFI byte will be 18 characters. When encoding according to the rules of URN Code 40, the overall length code with AFI byte will be 12 bytes. Thus, the total length of the entry in serial number field, together with the added byte value of the AFI application family, will be 104 bits. The resulting field size does not exceed the maximum possible size of the "serial number" field of the SGTIN-198, which is 140 bits.

This example is an illustration only and shows the principal possibility of placing a unique identifier of a library document in the structure of EPC code, which allows us to talk about integration of the integrated library systems and systems based on EPC standards. This possibility can be realized when using radio-frequency EPC labels in the supply chains with an EPC memory block capacity sufficient to accommodate the SGTIN-198 code. It will also require modification of the regulatory framework for the use of radio-frequency identification in libraries in terms of ISO 28560-4 standard, which defines rules for the use of specialized EPC RFID equipment in libraries. The possibility of realization of such extension of regulatory base of library RFID systems has ripened today and follows from the general logic of development of information systems and, in particular, library systems of radio-frequency identification.

7. The development of universal library HF/UHF RFID systems

In radio-frequency identification systems operating on the principles of EPC, there is currently no alternative to use equipment of UHF range, the work of which is defined by ISO/IEC 18000-63. The use of such equipment is fundamentally possible in library automation systems. This fact is reflected in ISO 28560 library standards. There are two standards defining operation of library RFID systems: ISO 28560-3—for HF systems, and ISO 28560-4—for UHF systems. The presence of two standards gives libraries an opportunity to choose the type of equipment for their automation system. Rules for placing and encoding data elements in the memory of radio-frequency labels defined by these standards are significantly different and are compatible only at the level of nomenclature of data elements presented in ISO 28560-1. Provisions given in the ISO 28560-2 are applicable only to RFID systems corresponding to the fourth part and are not applicable to systems based on the third part of ISO 28560. The reason for this is that RFID equipment of different frequency ranges—HF and UHF—is incompatible and that does these systems alternative to each other.

The use of UHF equipment based on the EPC concept in libraries is hampered by a number of factors; one of the most significant factors is the impossibility of a "smooth" migration from HF technology to UHF due to their complete incompatibility at the level of applied radio-frequency labels and readers. Thus, the choice of frequency range for the RFID library automation system uniquely determines the type of equipment that should be purchased by the library. The choice of equipment determines the overall configuration of the system and specialized software that is integrated into ILS. Subsequent migration to other equipment is practically impossible, because it is also associated with the re- or additional marking of library documents with radio-frequency labels of another type, since existing RFID readers are compatible only with labels of their frequency range.

The solution to this problem is possible by creating universal RFID systems that work with radio-frequency labels of both bands. Such a way requires a lot of efforts of software and hardware developers, which is associated with significant financial costs. Such costs are reasonable under the condition of payback due to widespread introduction of radio-frequency identification in a large number of libraries.

A common problem of widespread introduction of radio-frequency identification technology in libraries at present is the relatively high cost of equipment and availability only to libraries with good sources of finance. This is equally true for both HF and UHF systems. The cost of such equipment consists of two main components: the cost of RFID readers and the cost of radio-frequency tags. The share of radio-frequency tag cost in projects is growing rapidly with increasing library stock collection. For libraries with the collection of more than 100,000 document copies, to be labeled, the share of tags is already determining the cost of the RFID automation project.

Drastic cost reduction for the use of radio-frequency identification technology is possible due to repeated use of radio-frequency labels at several stages of product life cycle in the supply chain, from manufacturer to consumer. For libraries, this means that in order to reduce the cost of RFID library systems, printed publications must be marked with radio-frequency tags in manufacturer's printing house. Labels should be used to automate manufacturing, warehousing, and delivery processes. This is the main direction of development of automatic identification systems based on the EPC concept. Libraries will receive such documents already marked with EPC-type labels that carry information identifying a document as an object of accounting in the global EPC network. The use of such tags in RFID library systems will eliminate the need for libraries to purchase them independently, but the system must support work with tags, used in EPC systems. In other words, the library RFID system must be integrated into a global identification system, based on the EPC concept. Taking into account the fact that the EPC system is currently working with labels of UHF band, for most libraries with the book collection, marked with HF markers, such possibility will appear only in connection with appearance of universal systems that work with tags of both types.

The emergence of universal systems is dictated not only by needs of libraries. There are a number of areas, where use of HF radio-frequency labels is preferred due to physical properties of electromagnetic waves. Increased penetrability of working field of the HF readers allows more efficient reading of labels located inside the objects of accounting or inside package. Relatively small range of HF tags, which is essentially possible, makes them preferable in systems with high information security requirements. At the same time, inclusion of such systems in the global identification system significantly expands their functionality.

The universal systems are still a matter of the future, but their emergence is already supported by existing international standards. If we consider the recent history of the regulatory framework of radio-frequency identification technology over the last ~15 years, we can see a clear trend of transition of existing centers of standardization, from statement of current situation in the market of RFID equipment, to creation of a regulatory framework that determines and stimulates further development of technologies in the direction of integration of RFID systems of similar types.

Without going into the background history, we can say that in the beginning of 2000s, two standardization centers, working on basic RFID standards for logistics tasks, were formed in the world:

- Joint Technical Committee ISO/IEC JTC1 “Information technology” /SC 31 “Automatic identification and data capture techniques” developed by the group of standards ISO/IEC 18000 for all types of RFID equipment, including HF and UHF.

- “EPC global Inc.” organization is promoting the concept of electronic product code as a single identifier for all automatic identification systems, including RFID systems, and proposed a standard for manufacturers of UHF equipment “EPC UHF Class 1 Generation 2”.

The existence of two different standard groups that determined operation of similar types of devices and were incompatible with each other was a significant obstacle to development of RFID systems. None of them was fully supported by equipment manufacturers. From two ranges that are most used in practical fields, HF and UHF leading manufacturers of HF equipment (including library equipment) were guided by ISO/IEC 18000-3 Mode 1 standard and manufacturers of UHF equipment by EPC C1g2. Use of equipment of a particular range in specific areas was determined by their characteristics and limitations arising from physical properties of electromagnetic waves. In addition, the logical structure of labels of these ranges was significantly different from each other. This fact, along with the difference in frequency ranges, made HF and UHF systems alternative to each other.

The first step towards harmonization of two standardization trends was made by ISO/IEC JTC1. In 2006, a supplement was added to the existing ISO/IEC 18000 group of standards and an ISO/IEC 18000-63 standard was adopted, which defines a data exchange protocol between UHF RFID devices, compatible with the protocol EPC C1g2.

The next step was the development in 2011, by the GS1 international organization together with EPC global Inc., of a standard "EPC Class 1 HF", which defined EPC protocol concepts for HF RFID equipment. The new standard was supported by ISO/IEC JTC1/SC31 by adopting a similar supplement "Mode 3" to ISO/IEC 18000-3 standard.

At present, we can talk about the existence of a regulatory framework for production and use of RFID equipment, specialized for operation in automation systems, based on the concept of EPC and operating in both frequency bands [22]. In this case, RFID readers of both types will have the same functionality, and radio-frequency tags will have a similar logical structure, described in the following international standards:

- HF band—ISO/IEC 18000-3 Mode3 (EPC Class 1 HF)
- UHF band–ISO/IEC 18000-63 Type C (EPC C1g2)

The existence of a common regulatory framework for production and use of RFID equipment in the most popular frequency bands provides a fundamental opportunity to implement the original EPC concept: using a single electronic product code to identify objects of accounting in RFID systems of various specializations, including library.

The ability of "transparent" work of the EPC RFID system in two ranges, along with the use of EPC tags, requires use of double-frequency RFID tag readers. Creation of such tag readers is a highly technical task. The first step in this direction was made by FEIG Electronic company, which began production of "ID ISC. PRHD102 HF/UHF" [23] mobile readers in 2013, which supported simultaneous operation in HF/UHF bands. The RFID reader cost of this model is approximately 20% higher than the cost of a similar UHF range reader.

Despite the presence of a dual band RFID reader, its use in the proposed generic library system today is not possible, as it does not support working with HF tags like ISO/IEC 18000-3 Mode 3 (EPC Class 1 HF). In addition, until today it is the only dual-band RFID reader in the market, and readers of various types are required to create RFID library systems.

The possibility of producing label EPC HF band appeared in 2013, when NXP Semiconductors company started production of ICODE ILT type chips [24], which comply with ISO/IEC 18000-3 Mode 3 (EPC Class 1 HF). On the basis of these chips it is possible to produce library HF labels of EPC type, but until today such labels are not presented in the market and are not used in RFID systems.

From the aforesaid, it is visible that emergence of universal RFID library systems today is problematic. Developers of RFID systems face the problem of lack of necessary equipment in the market: tags and readers, and manufacturers are in no hurry to invest in establishing production of new equipment due to the lack of created sales market. Current situation remained very similar to that of the 1990ths for the whole radio-frequency identification technology.

Implementation of the large RFID system project, with the use of HF band EPC tags, can change the situation in an area where use of this band labels is advisable along with UHF labels. Within the framework of such a project, developments can be made, further commercialization of which can change the situation in the market. Implementation of such project at the level that ensures its economic efficiency is possible only for large commercial or state organization. RFID library systems occupy a very modest place in the total number of RFID systems and unlikely will be able to meet their needs of the required scale project.

Participation of libraries in the overall development of automatic identification systems may be in their integration into the supply chain of printed publications, from publisher to reader, along with trading organizations. To do this, RFID library systems must support the EPC concept, and it will make a notable contribution to the development of this concept. Such support requires the development and expansion of the regulatory framework that defines principles of application of RFID equipment, designed for EPC systems, in libraries.

Participation in development of the global EPC network will be useful for libraries. The fundamental difference between libraries and book-selling organizations is that documents of library collection are transferred to users for a limited period, with their subsequent return to the library storage system. At the same time, libraries provide users with advanced opportunities to search for necessary information. Integration of ILS into the global EPC identification system, using services of the Trusted Source of Data network can significantly expand the search capabilities of ILS not only for users of library services, but also for library acquisition services. Inclusion of marked documents of library collections into information space of automated systems, based on EPC standards, can significantly increase their mobility in delivery services of the ILL system in the future to provide the widest possible range for access to the library's holdings through widespread use of new information technologies, with using technologies of automatic identification and item management.

8. Integration of library RFID systems into the network of Internet of things

At present, the Internet of Things (IoT) concept is actively developing. This concept involves creation of a computer network that combines physical objects equipped with means to interact with each other and with ambient medium. Great importance in this network is given to artificial intelligence systems, managing processes and excluding human participation from certain actions and operations. Establishment of such a network is possible only on the basis of standardization of information exchange principles. An important place in development of the IoT concept is taken by technologies of automatic identification, and among them the technology of radio-frequency identification has a leading value. The concept of EPC as the global identification system participates in general development of the Internet of Things [25].

IOT network can be represented as a virtual space consisting of objects identified in a standard way, and there are used standard communication channels. Within the frame of such a surrounding, there may be many functionally localized information systems that interact "transparently" with each other. Library documents may participate in such systems. This can significantly extend the functionality of library automation systems. The development of IoT systems with library functionality may partially replace functions of specialized library automation systems. The central place in localized IoT systems is taken by humans, who determine the purpose of activity in any area. The system actively involves information for the program to achieve this purpose and making conditions for its implementation. Information support is a key condition for any kind of human activity. Participation of systems of the IoT in information support means that the system itself will define information needs of people, to select and provide sources of necessary information which are in the area accessible for them.

Such an area is the global information system which has included information objects identified in a standard way. These objects may be electronic documents and printing editions marked by radio-frequency tags identified in the global system of identification. The system can determine nearest location of available copy of desired edition independently, and even it can order its delivery to the required place as the location of necessary publications can be both, trading organizations and libraries. For libraries, this will mean that the number of users of their information services, along with a person, will include expert systems of artificial intelligence. This requires that library document collections exist in the global system of automatic identification and in information space of the IoT network.

The development of IoT as a new communication technology is very fast. Today, such systems are already widely used in automation systems of mass production as Industrial Internet of Things. Mass appearance of such systems in the consumer sphere is predicted in 5–10 years. Such systems already exist in space of electronic information resources in the Internet environment. These are the so-called WEB 3.0 systems, the concept of which was formed in the mid of 2000s [26]. Now they exist as technological platforms for formation of content of the smart websites. Inclusion of physical objects into operational space of such systems, which may include library documents, will mean exit of WEB 3.0 systems from Internet virtual space to the real world and transition to IoT systems. This transition is directly related to automatic identification technologies and, to a large extent, to development of radio-frequency identification technology.

The development of library systems of radio-frequency identification in the direction of the EPC and the Internet of things concepts will allow including traditional printed documents, which make up a large part of the library stock collections, in digital information space, along with electronic documents. This will increase accessibility of printed documents to users and promote development of library technologies. This will allow libraries to integrate more fully into global information space at the next stage of development of information technologies as a global integrated system of library and information support of human activities and take a worthy place in the modern information society.

9. Conclusion

The appearance of radio-frequency identification technology is associated with the development of microelectronics and computer technology. This technology is also the general direction of automatic identification technology development, which allowed effective use of the computer technology in a wide range of applications. The use of first bar-code and then radio-frequency identification in libraries has significantly improved the traditional methods of servicing readers.

The emergence of electronic documents was the next step in the development of information technology, which allowed libraries to go beyond reading rooms to the limitless expanses of the Internet. The new opportunities have become a serious challenge for traditional documents of library collections, which are significantly inferior to electronic documents in access speed. There were ideas of a total elimination of paper books, but such forecasts do not sound today. The market for paper books has been growing in recent years, which means that paper books have found their place in the modern world. Increasing the availability of paper books in the electronic information space is a very urgent task today.

Printed documents now are a significant part of the library collections around the world. The use of radio-frequency identification technology in libraries together with the concept of the Internet of things will allow including “traditional” printed documents in the digital information space along with electronic documents. This will contribute to the development of library technologies and will allow more fully integrate libraries in the global information space at the next stage of development of information technologies.

Author details

Timoshenko Igor
Russian National Public Library for Science and Technology, Moscow, Russian Federation

*Address all correspondence to: timigor@gpntb.ru

References

[1] Paul Pandian M. RFID for Libraries. A Practical Guide. Chandos Publishing; 2010. 186 p. ISBN: 9781843345459. eBook ISBN: 9781780630489. Available from: https://www.kobo.com/us/en/ebook/rfid-for-libraries

[2] Timoshenko IV. The application of the radio frequency identification technology at libraries: The Russian experience. Scientific and Technical Information Processing. 2016;**43**(3):189-193. DOI: 10.3103/S0147688216030126

[3] Kosareva E. Introducing RFID Technology in the Library of St. Petersburg State University Graduate School of Management: Eleventh International Conference and Exibition LIBCOM-2007; 12-16 November 2007; Zvenigorod. Moscow; 2007. CD-ROM (ISBN 978-5-85638-120-6). p. 15

[4] Abramov SB, Vishnitskii AF, Reznik AY, Timoshenko IV. Sistema avtomatizatsii raboty biblioteki na osnove tekhnologii RFID. Informatsionnye tekhnologii i sistemy: (The system for automation of the library operation based on the RFID Technology). In: Barinov V, editor. Information Technologies and Systems: Interuniversity Collection of Scientific Papers. Moscow: MIET; 2009. pp. 212-216

[5] RFID Data Model for Libraries. Danish Standard S24/u4. 2005. Available from: http://biblstandard.dk/rfid/dk/RFID_Data_Model_for_Libraries_July_2005.pdf [Accessed: September 06, 2018]

[6] ISO 28560-1:2014. Information and Documentation—RFID in Libraries—Part 1: Data Elements and General Guidelines for Implementation. Available from: https://www.iso.org/standard/65203.html

[7] ISO 15511:2011. Information and Documentation—International Standard Identifier for Libraries and Related Organizations (ISIL). Available from: https://www.iso.org/standard/57332.html

[8] ISO 20247:2018. Information and Documentation—International Library Item Identifier (ILII). Available from: https://www.iso.org/standard/67408.html

[9] ISO 28560-2:2014. Information and Documentation—RFID in Libraries—Part 2: Encoding of RFID Data Elements Based on Rules From ISO/IEC 15962. Available from: https://www.iso.org/standard/65204.html

[10] ISO 28560-3:2014. Information and Documentation—RFID in Libraries—Part 3: Fixed Length Encoding. Available from: https://www.iso.org/standard/65205.html

[11] ISO/IEC 18000-3:2010. Information Technology—Radio Frequency Identification for Item Management—Part 3: Parameters for Air Interface Communications at 13.56 MHz. Available from: https://www.iso.org/standard/53424.html

[12] ISO/TS 28560-4:2014. Information and Documentation—RFID in Libraries—Part 4: Encoding of Data Elements Based on Rules From ISO/IEC 15962 in an RFID Tag With Partitioned Memory. Available from: https://www.iso.org/standard/62311.html

[13] ISO/IEC 18000-63:2015. Information Technology—Radio Frequency Identification for Item Management—Part 63: Parameters for Air Interface Communications at 860 MHz to 960 MHz Type C. Available from: https://www.iso.org/standard/63675.html

[14] EPC™ Radio-Frequency Identity Protocols. Generation-2 UHF RFID Standard. Specification for RFID Air Interface Protocol for Communications at 860 MHz–960 MHz. Release 2.1,

Ratified. July 2018. Available from: https://www.gs1.org/sites/default/files/docs/epc/GS1-EPC-Gen2v2-HF-AirInterface_i21_r_2018-09-04.pdf

[15] ISO 27730:2012. Information and Documentation—International Standard Collection Identifier (ISCI). Available from: https://www.iso.org/standard/44293.html

[16] ISO 3166-1:2013. Codes for the Representation of Names of Countries and Their Subdivisions—Part 1: Country Codes. Available from: https://www.iso.org/standard/63545.html

[17] ISO/IEC FDIS 15961-2. Information Technology—Radio Frequency Identification (RFID) for Item Management: Data Protocol—Part 2: Registration of RFID Data Constructs. Available from: https://www.iso.org/standard/43631.html

[18] ISO/IEC FDIS 15961-3. Information Technology—Radio Frequency Identification (RFID) for Item Management: Data Protocol—Part 3: RFID Data Constructs. Available from: https://www.iso.org/standard/43632.html

[19] Global Trade Item Number (GTIN) [Internet]. Available from: https://www.gs1.org/standards/id-keys/gtin [Accessed: September 6, 2018]

[20] EPC™ Radio-Frequency Identity Protocols. EPC Class-1 HF RFID Air Interface Protocol for Communications at 13.56 MHz. Version 2.0.3. 2011. Available from: https://www.gs1.org/sites/default/files/docs/epc/epcglobal_hf_2_0_3-standard-20110905r3.pdf

[21] EPC Tag Data Standard defines the Electronic Product Code™ and Specifies the Memory Contents of Gen 2 RFID Tags. Release 1.11, Ratified. 2017. Available from: https://www.gs1.org/sites/default/files/docs/epc/GS1_EPC_TDS_i1_11.pdf

[22] Timoshenko IV. Radio-frequency identification technology in libraries. Integration of library management systems into global identification systems. Scientific and Technical Information Processing. 2017;**44**(4):280-284. DOI: 10.3103/S0147 688217040116

[23] HF/UHF HAND-HELD READER [Internet]. Available from: https://www.feig.de/en/products/identification/product/id-iscprhd102/ [Accessed: September 6, 2018]

[24] ICODE ILT-M.Rev. 3.2—8 October 2013. Product data sheet. Available from: https://www.nxp.com/docs/en/data-sheet/SL2S1412_SL2S1512_SL2S1612.pdf

[25] ISO/IEC 29161:2016. Information Technology—Data Structure—Unique Identification for the Internet of Things. Available from: https://www.iso.org/standard/45240.html

[26] Is Web 3.0 Really a Thing? A Brief Intro to Web 3.0 and What to Expect [Internet]. Available from: https://www.lifewire.com/what-is-web-3-0-3486623 [Accessed: September 6, 2018]